Causality and Chance in Modern Physics

Causality and Chance in Modern Physics
David Bohm

Foreword by Louis De Broglie

PENN

UNIVERSITY OF PENNSYLVANIA PRESS
PHILADELPHIA

Originally published 1957 by Routledge & Kegan Paul and D. Van Nostrand
Company, Inc.; reprinted 1961 by Harper & Row Publishers.
Copyright © 1957 David Bohm

Printed in the United States of America on acid-free paper

10 9 8 7

Published by
University of Pennsylvania Press
Philadelphia, Pennsylvania 19104-4011

Library of Congress Cataloging-in-Publication Data

Bohm, David
    Casuality and Chance in Modern Physics / David Bohm; foreword by
Louis de Broglie
    Includes bibliographical references

ISBN 0-8122-1002-6
1. Physics—philosophy. I. Title. II. de Broglie, Louis

QC6.B597   57-28894
530.1   CIP

CONTENTS

FOREWORD BY LOUIS DE BROGLIE   *page* ix

I CAUSALITY AND CHANCE IN NATURAL LAW

1 Introduction	1
2 Causality in Natural Processes	4
3 Association *v.* Causal Connection	5
4 Significant Causes in a Given Context	7
5 More General Criteria for Causal Relationships	10
6 Causal Laws and the Properties of Things	12
7 One-to-Many and Many-to-One Causal Relationships	16
8 Contingency, Chance, and Statistical Law	20
9 The Theory of Probability	25
10 General Considerations on the Laws of Nature	28
11 Conclusion	32

II CAUSALITY AND CHANCE IN CLASSICAL PHYSICS: THE PHILOSOPHY OF MECHANISM

1 Introduction	34
2 Classical Mechanics	34
3 The Philosophy of Mechanism	36
4 Developments away from Mechanism in Classical Physics	39
5 Wave Theory of Light	40
6 Field Theory	41
7 On the Question of What is the Nature of the Electromagnetic Field	43
8 Field Theories and Mechanism	45
9 Molecular Theory of Heat and the Kinetic Theory of Gases	47

Contents

10 On the Relationship between Microscopic and
Macroscopic Level, according to the Molecular
Theory   49
11 Qualitative and Quantitative Chances   52
12 Chance, Statistical Law, and Probability in Physics   54
13 The Enrichments of the Conceptual Structure of
Classical Physics and the Philosophy of
Mechanism   56
14 A New Point of View towards Probability and
Statistical Law—Indeterministic Mechanism   62
15 Summary of Mechanism   65

III THE QUANTUM THEORY

1 Introduction   68
2 Origin of the Quantum Theory   70
3 On the Problem of Finding a Causal Explanation
of the Quantum Theory   79
4 The Indeterminacy Principle   81
5 Renunciation of Causality in Connection with the
Atomic Domain a Consequence of the Indeterminacy Principle   84
6 Renunciation of Concept of Continuity in the
Atomic Domain   89
7 Renunciation of all well-defined Conceptual Models
in the Microscopic Domain—The Principle of
Complementarity   91
8 Criticism of Conclusions drawn in the usual Interpretation of the Quantum Theory on the Basis of
the Indeterminacy Principle   94
9 The Usual Interpretation of the Quantum Theory
a Form of Indeterministic Mechanism   101

IV ALTERNATIVE INTERPRETATIONS OF THE QUANTUM THEORY

1 Introduction   104
2 General Considerations on the Sub-Quantum
Mechanical Level   106
3 Brief Historical Survey of Proposals for Alternative
Interpretations of the Quantum Theory   109
4 A Specific Example of an Alternative Interpretation of the Quantum Theory   111

Contents

5 Criticisms of Suggested New Interpretation of
the Quantum Theory 117
6 Further Developments of the Theory 118
7 The Current Crisis in Microscopic Physics 121
8 Advantages of New Interpretation of Quantum
Theory in the Guidance of Research in New
Domains 123
9 Alternative Interpretation of the Quantum Theory
and the Philosophy of Mechanism 126
Bibliography 128

V MORE GENERAL CONCEPT OF NATURAL LAW

1 Introduction 130
2 Summary of the Essential Characteristics of a
Mechanistic Philosophy 130
3 Criticism of the Philosophy of Mechanism 131
4 A Point of View that goes beyond Mechanism 132
5 More Detailed Exposition of the Meaning of
Qualitative Infinity of Nature 137
6 Chance and Necessary Causal Interconnections 140
7 Reciprocal Relationships and the Approximate and
Relative Character of the Autonomy of the Modes
of being of Things 143
8 The Process of Becoming 146
9 On the Abstract Character of the Notion of
Definite and Unvarying Modes of Being 153
10 Reasons for Inadequacy of Laplacian Determinism 158
11 Reversibility versus Irreversibility of the Laws of
Nature 160
12 Absolute versus Relative Truth—The Nature of
Objective Reality 164

FOREWORD

By Louis de Broglie

THOSE who have studied the development of modern physics know that the progress of our knowledge of microphysical phenomena has led them to adopt in their theoretical interpretation of these phenomena an entirely different attitude to that of classical physics. Whereas with the latter, it was possible to describe the course of natural events as evolving in accordance with causality in the framework of space and time (or relativistic space-time), and thus to present clear and precise models to the physicist's imagination, quantum physics at present prevents any representations of this type and makes them quite impossible. It allows no more than theories based on purely abstract formulæ, discrediting the idea of a causal evolution of atomic and corpuscular phenomena; it provides no more than laws of probability: it considers these laws of probability as having a primary character and constituting the ultimate knowable reality: it does not permit them to be explained as resulting from a causal evolution which works at a still deeper level in the physical world.

We can reasonably accept that the attitude adopted for nearly 30 years by theoretical quantum physicists is, at least *in appearance*, the exact counterpart of information which experiment has given us of the atomic world. At the level now reached by research in microphysics it is certain that the methods of measurement do not allow us to determine simultaneously all the magnitudes which would be necessary to obtain a picture of the classical type of corpuscles (this can be deduced from Heisenberg's uncertainty principle), and that the perturbations introduced by the measurement, which are impossible to eliminate, prevent us in general from predicting precisely the result which it will produce and allow only statistical predictions. The construction of purely probablistic formulæ that all theoreticians use today was thus completely justified. However, the majority of them, often under the influence of preconceived ideas derived from positivist doctrine, have thought that they could go further and assert that the uncertain and incomplete character of the knowledge that experiment at its present stage gives us about what really happens in microphysics is the result of a real indeterminacy of the physical states and of their evolution. Such an extrapolation does not appear in any way to be justified. It is possible that looking into the future to a deeper level of physical reality we will be able to interpret the laws of probability and quantum physics as being the statistical results of the development of completely determined values of variables which are at present hidden from us. It may be that the powerful means we are beginning to use to break up the structure of the nucleus and to make new particles appear will give us one day a direct knowledge which we do not now have of this deeper level. To try to stop all attempts to pass beyond the present viewpoint of quantum physics could be very dangerous for the progress of science and would furthermore be contrary to the lessons we may learn from the history of science. This teaches us, in effect, that the actual state of our knowledge is always provisional and that there must be, beyond what is actually known, immense new regions to discover. Besides, quantum physics has found itself for several years tackling problems which it has not been able to solve and seems to have arrived at a dead end. This situation suggests strongly that an effort to modify the framework of ideas in which quantum physics has voluntarily wrapped itself would be valuable.

One is glad to see that in the last few years there has been a development towards re-examining the basis of the present interpretation of microphysics. The starting point of this movement was two articles published at the beginning of 1952 by David Bohm in the *Physical Review*. A long time ago in an article in the *Journal de Physique* of May 1927 I put forward a causal interpretation of wave mechanics which I called the "theory of double solutions" but I abandoned it, discouraged by criticisms which this attempt roused. In his 1952 paper Professor Bohm has taken up certain ideas from this article and commenting and enlarging on them in a most interesting way he has successfully developed important arguments in favour of a causal reinterpretation of quantum physics. Professor Bohm's paper has led me to take my old concepts up again, and with my young colleagues at the Institute, Henri Poincaré, and in particular M. Jean-Pierre Vigier, we have been able to obtain certain encouraging results. M. Vigier working with Professor Bohm himself has developed an interesting interpretation of the statistical significance of $|\psi|^2$ in wave mechanics. It seems desirable that in the next few years efforts should continue to be made in this direction. One can, it seems to me, hope that these efforts will be fruitful and will help to rescue quantum physics from the cul-de-sac where it is at the moment.

In order to show the legitimacy and also the necessity of such attempts, Professor Bohm has thought that the moment had come to take up again in his researches the critical examination of the nature of physical theories and of interpretations which are susceptible to explaining natural phenomena as fast as science progresses. He has compared the development of classical physics, where in succession the viewpoint of universal mechanism, and then of the general theory of fields, and then of statistical theories have appeared, one after the other, with the introduction by quantum physics of its own new conceptions. He has shrewdly and carefully analysed the idea of chance and has shown that it comes in at each stage in the progress of our knowledge, when we are not aware that we are at the brink of a deeper level of reality, which still eludes us. Convinced that theoretical physics has always led, and will always lead, to the discovery of deeper and deeper levels of the physical world, and that this process will continue without any limit, he has concluded that quantum physics has no right to consider its present concepts definitive, and that it cannot stop researchers imagining deeper domains of reality than those which it has already explored.

I cannot give here a complete account of the thorough and fascinating study which Professor Bohm has made. The reader will find a very elegant and suggestive analysis which will instruct him and make him think. No one is better qualified than Professor Bohm to write such a book, and it comes exactly at the right time.

CHAPTER ONE

Causality and Chance in Natural Law

1. INTRODUCTION

IN nature nothing remains constant. Everything is in a perpetual state of transformation, motion, and change. However, we discover that nothing simply surges up out of nothing without having antecedents that existed before. Likewise, nothing ever disappears without a trace, in the sense that it gives rise to absolutely nothing existing at later times. This general characteristic of the world can be expressed in terms of a principle which summarizes an enormous domain of different kinds of experience and which has never yet been contradicted in any observation or experiment, scientific or otherwise; namely, everything comes from other things and gives rise to other things.

This principle is not yet a statement of the existence of causality in nature. Indeed, it is even more fundamental than is causality, for it is at the foundation of the possibility of our understanding nature in a rational way.

To come to causality, the next step is then to note that as we study processes taking place under a wide range of conditions, we discover that inside of all of the complexity of change and transformation there are *relationships* that remain effectively constant. Thus, objects released in mid-air under a wide range of conditions quite consistently fall to the ground. A closer study of the rate of fall shows that in so far as air resistance can be neglected, the acceleration is constant; while still more general relationships can be found that hold when air resistance has to be taken into account. Similarly, water put into a container quite invariably "seeks its own level" in a wide range of conditions. Examples of this kind can be multiplied without limit. From the extreme generality of this type of behaviour, one begins to consider the possibility that in the processes by which one thing comes out of others, the constancy of certain relationships inside a wide variety of transformations and changes is no coincidence. Rather, we interpret this constancy as signifying that such

1

relationships are *necessary*, in the sense that they could not be otherwise, because they are inherent and essential aspects of what things are. The necessary relationships between objects, events, conditions, or other things at a given time and those at later times are then termed *causal laws*.

At this point, however, we meet a new problem. For the necessity of a causal law is never absolute. For example, let us consider the law that an object released in mid-air will fall. This in fact is usually what happens. But if the object is a piece of paper, and if "by chance" there is a strong breeze blowing, it may rise. Thus, we see that one must conceive of the law of nature as necessary only if one abstracts* from *contingencies*, representing essentially independent factors which may exist outside the scope of things that can be treated by the laws under consideration, and which do not follow necessarily from anything that may be specified under the context of these laws. Such contingencies lead to *chance*.† Hence, we conceive of the necessity of a law of nature as *conditional*, since it applies only to the extent that these contingencies may be neglected. In many cases, they are indeed negligible. For example, in the motion of the planets, contingencies are quite unimportant for all practical purposes. But in most other applications, contingency is evidently much more important. Even where contingencies are important, however, one may *abstractly* regard the causal law as something that *would* apply if the contingencies were not acting. Very often we may for practical purposes isolate the process in which we are interested from contingencies with the aid of suitable experimental apparatus and thus verify that such an abstract concept of the necessity of the causal relationships is a correct one.

Now, here it may be objected that if one took into account *everything* in the universe, then the category of contingency would disappear, and all that happens would be seen to follow necessarily and inevitably. On the other hand, there is no known causal law that really does this. It is true that in any given problem we may, by

---

* Throughout this book, we shall use the word "abstract" in its literat sense of "taking out". When one abstracts something, one simplifies il by *conceptually* taking it out of its full context. Usually, this is done by taking out what is common to a wide variety of similar things. Thus, abstractions tend to have a certain generality. Whether a particular abstraction is valid in a given situation then depends on the extent to which those factors that it ignores do in fact produce negligible effects in the problems of interest.

† We are here taking the word "contingency" in its widest sense; namely, the opposite of necessity. Thus, contingency is that which could be otherwise. Chance will then later be seen to be a certain very common form of contingency, while causality will likewise be seen to be a special but very common form of necessity.

broadening the context of the processes under consideration, even find the laws which govern some of the contingencies. Thus, in the case of the piece of paper being blown around by the wind, we could eventually study the laws which determine how the wind will blow. But here we will meet new contingencies. For the behaviour of the wind depends on the locations of the clouds, on the temperatures of bodies of water and land, and even as shown in some of the latest meteorological studies, on beams of electrons and ultraviolet rays which may be emitted with unusual intensity during sunspots. This means, however, that we must now go into the laws governing the formation of clouds, of land masses, of bodies of water, and of the processes in which the sunspots originate. Thus far, no evidence has been discovered that the possibility of tracing causal relationship in this way will ever end. In other words, every real causal relationship, which necessarily operates in a finite context, has been found to be subject to contingencies arising outside the context in question.*

To understand the relationship between causality and contingency that has actually been found thus far, we may compare these two categories to two opposite views of the same object. Each view is an abstraction which by itself gives an adequate idea of certain aspects of the object, but which will lead to erroneous results if we forget that it is, after all, only a partial view. Each view, then, limits the other, corrects the other, and through its relationship with the other enables us to form a better concept of what the object is. Of course, we may take an infinity of different views, but associated with each view there is always an opposite view. Thus, while we can always view any given process from any desired side (e.g. the causal side) by going to a suitable context, it is always possible to find another context in which we view it from the opposite side (in this case, that of contingency).

In sum, then, we may say that the processes taking place in nature have been found to satisfy laws that are more general than those of causality. For these processes may also satisfy *laws of chance* (which we shall discuss in more detail in Sections 8 and 9), and also laws which deal with the relationships between causality and chance. The general category of law, which includes the causal laws, the laws of chance, and the laws relating these two classes of law, we shall call by the name of *laws of nature*.

* Various purely philosophical efforts to define causal laws that are completely free of contingency have been made. Such efforts are based on a *mechanistic* point of view towards the world. The inadequacy of this point of view will be made clear in Chapter II and in Chapter V.

2. CAUSALITY IN NATURAL PROCESSES

The causal laws in a specific problem cannot be known *a priori*; they must be *found* in nature. However, in response to scientific experience over many generations along with a general background of common human experience over countless centuries, there have evolved fairly well-defined methods for finding these causal laws. The first thing that suggests causal laws is, of course, the existence of a regular relationship that holds within a wide range of variations of conditions. When we find such regularities, we do not suppose that they have arisen in an arbitrary, capricious, or coincidental fashion, but, as pointed out in the previous section, we assume, at least provisionally, that they are the result of necessary causal relationships. And even with regard to the irregularities, which always exist along with the regularities, one is led on the basis of general scientific experience to expect that phenomena that may seem completely irregular to us in the context of a particular stage of development of our understanding will later be seen to contain more subtle types of regularity, which will in turn suggest the existence of still deeper causal relationships.

Having found some regularities which we provisionally suppose are the results of causal laws, we then proceed to make *hypotheses* concerning these laws, which would explain these regularities and permit us to understand their origin in a rational way.* These hypotheses will in general lead to new *predictions*, of things not contained in the empirical data which gave rise to them. Such predictions may then be *tested*, either by simple observation of phenomena that take place of their own accord, or by the more active procedure of doing an experiment, or of applying the hypotheses as a guide in practical activities.

In observations and experiments, an effort is made to choose conditions in which the processes of interest are isolated from the interference of contingencies. Although no such effort can lead to a complete avoidance of contingencies, it is often possible to obtain a degree of isolation that is good enough for practical purposes. If, then, the predictions based on our hypotheses are consistently verified in a wide range of conditions, and if, within the degree of approximation with which we are working, all failures of verification can be understood as the results of contingencies that it

---

* By explanation, of a given thing, one means the demonstration that this thing follows necessarily from other things. An explanation therefore reduces the number of arbitrary elements in any given context.

was not possible to avoid,* then the hypothesis in question is accepted as an essentially correct one, which applies at least within the domain of phenomena that have been studied, as well as very probably in many new domains that have not yet been studied. If such a verification is not obtained, then it is of course necessary to go back and to seek new hypotheses until it has been obtained.

Even after correct hypotheses have been developed, however, the process does not stop here. For such hypotheses will, in general, lead to new observations and experiments, and to new kinds of practical activities, out of which may come the discovery of new empirical regularities, which in turn require new explanations, either in terms of a modification of existing hypotheses or in terms of a fundamental revision of one or more of the hypotheses underlying these hypotheses. Thus, theoretical explanations and empirical verifications each complement and stimulate the other, and lead to a continual growth and evolution of science, both with regard to theory and with regard to practice and to experiment.

It is necessary, however, to make the presentation of causality given in this section more precise. This we shall now proceed to do with the aid of a wide range of examples, which show how various aspects of causal relationships actually manifest themselves in specific cases.

3. ASSOCIATION V. CAUSAL CONNECTION

The first problem that we shall consider is to analyse more carefully the relationship between causality and a regular association of conditions or events. For a regular association between a given set, A, of events or conditions in the past, and another set, B, in the future does not necessarily imply that A is the cause of B. Instead, it may imply that A and B are associated merely because they are both the result of some common set of causes, C, which is anterior to both A and B. For example, before winter the leaves generally fall off the trees. Yet the loss of the leaves by the trees is not the cause of winter, but is instead the *effect* of the general process of lowering of temperature which first leads to the loss of leaves by the trees and later to the coming of winter. Clearly, then, the concept of a causal relationship implies more than just regular association, in which one set of events precedes another in the time. What is implied in addition is that (abstracted from contingencies, of course) the future effects come out of past causes through a process satisfying *necessary*

---

* E.g. when we see a piece of paper in mid-air that is not falling, we must find that something is happening (for example a breeze is blowing) which accounts for the failure of our prediction that objects released in the earth's gravitational field will fall towards the earth.

relationships. And, as is evident, mere association is not enough to prove this kind of connection.

An important way of obtaining evidence in favour of the assumption that a given set of events or conditions comes necessarily from another is to show that a wide range of *changes* in one or more of the presumed causes occurring under conditions in which other factors are held constant always produces corresponding changes in the effects. The more co-ordinations of this kind that one can demonstrate in the changes of the two sets of events, the stronger is the evidence that they are causally related; and with a large enough number one becomes, for practical purposes, certain that this hypotheses of causal connection is correct. To obtain such a demonstration, however, an active interference on our part by means of experiments will usually be required, although in some cases enough changes of the right kind will occur naturally so that it will be adequate to make a wide range of observations in the phenomena that are already at hand.

We may illustrate how suitable experiments and observations make possible a distinction between a regular association of events and causality by means of an example taken from the field of medicine. Originally, it was noticed that the disease malaria was associated with the damp air of night. Thus, it was thought that the damp night air was the cause of malaria. But this hypothesis did not explain the known facts very well. For it was found that malaria could exist even in places where the air was dry, while it was often absent in places where the air was very damp. But it was noted that in places where the night air was damp, there were many mosquitoes, which could bite people who left their windows open. The hypothesis then considered was that the mosquito carried something from the blood of a sick person to the blood of a healthy person, which could cause malaria. Such a hypothesis could explain why malaria was generally found in damp places, since in such places there are many mosquitoes. It also explained why malaria could be produced even in dry places, so long as there were occasional pools in which mosquitoes could breed. Finally, it explained why damp places could exist without malaria, provided that there were no people with the disease in the neighbourhood. Thus, a hypothesis had been produced which could explain a wide range of facts, at least in a general sort of way. To verify this hypothesis, however, experiments were needed, specially designed in order to eliminate the possibility that mosquitoes were only regularly associated with the disease, while damp air would be one of the real causes. Various volunteers were taken, and divided into three groups. All groups were isolated to prevent bites by mosquitoes that may have been in the neighbourhood by chance. The first group was not allowed to be bitten by mosquitoes at all, the second was bitten only by mosquitoes which had no access to people with malaria, and the third was bitten by mosquitoes that had bitten people having malaria. All three groups were divided into two parts: one part exposed to damp air, the other part not. Only those in the third group caught malaria, and of these, only those who had been bitten by a special type of mosquito (Anopheles). The change between damp and dry air made no difference in any of the groups, thus showing that this factor had been a mere association* and not a true cause. On the other hand, the elimination of the Anopheles mosquitoes or the lack of contact with people who were infected with malaria eliminated the disease. The true cause, therefore, had to be something transmitted by the Anopheles mosquito from the blood of a sick person to the blood of a healthy person. Later work showed that this something is a definite bacterium.

This example shows the value of controlled experiments in distinguishing a true cause from an irrelevant association. It also shows how a search for an improved explanation of the facts will often help disclose some of the true causes. Finally, it shows· the importance of discovering such a cause; for this discovery made possible the control of malaria, as well as aiding in the search for remedies which would kill the malaria-producing bacterium.

4. SIGNIFICANT CAUSES IN A GIVEN CONTEXT

We have simplified the problem considerably in the previous example, by supposing that there is only one cause of malaria. In reality, the problem is much more complex than has been indicated. For not everybody who is bitten by an infected mosquito gets sick. This fact is explained by a more detailed understanding of the processes involved in getting sick. Thus, the bacteria produce substances that interfere with the functioning of the body and tend to make a person sick. But the body can produce substances which interfere with the functioning of the bacteria. Thus, two opposing tendencies are set up. Which one will win depends on complex factors concerning the functioning of microbes and of the body, which are not yet fully understood. But we see that it is too simple to think of the microbe as the *only* cause of malaria. Actually, it merely tends to initiate the processes which lead to sickness, and thus merely contributes to the production of malaria.

But now, if we admit the idea that each condition or event has

* It is clear that the damp air and the growth of mosquitoes generally have a common cause (i.e. bodies of stagnant water), which explains why they are frequently associated.

many contributing causes, we are led to a series of new problems. First of all we note that all events and objects in the universe have thus far shown themselves to be interconnected in some way even if perhaps only slightly. Strictly speaking, then, one should say that everything may have an infinite number of contributing causes. But in practice most of these have a negligible effect in the problem of interest. Thus we may define the "significant causes" of a given effect as those conditions or events which, in the context of interest, have appreciable influence on the effects in question.

As an example, consider the problem of malaria again. Now the moon exerts a gravitational force on every object in the universe, and therefore it must have an influence both on the malaria bacterium and on the person who might get malaria. In practice, however, this influence is usually negligible. But not always. For the moon can raise tides, which can push back a stream that flows into the sea, and thus create fresh water pools in which mosquitoes might breed. In certain cases, therefore, the moon could be an indirect contributing cause of malaria. Hence, the question of what are the "significant causes" in any particular problem cannot be solved *a priori*, but must in general be decided in each case only after a careful study, with the object of finding the factors that are necessary in the context of interest for the production of the essential features of the effect in question.

Even after we have settled which factors may be neglected, serious problems remain for us to solve. One of those is that of knowing when we have included *all* of the significant causes. For the mere proof that a change of the presumed cause has an appreciable influence on the effect when other presumed causes are held constant shows only that we have discovered *one* of the significant causes. As a means of indicating at least when we have *failed* to discover all of the significant causes, there has evolved the test of reproducibility. This test is based on the principle that if we reproduce *all* of the significant causes, then the effect must be reproduced at least in its essential aspects. Thus, a discovery that the results of an experiment are not reproducible suggests that one or more of the significant causes are varying from one experiment to the next, and thus producing a variation in the effect. This is essentially an application of the principle introduced at the beginning of this chapter; namely, that everything comes from something else. Thus, in this case, we do not admit the possibility of arbitrary variations of an effect that are totally unrelated to variations in the state of the things from which the effect came. If unexplained variations in the effect are found, it is then necessary to discover, by means of carefully controlled experiments guided by hypotheses based on the available facts, what is responsible for the lack of reproducibility of the effects. For example, in the case of the disease malaria, we have already cited the fact that the bite of an infected mosquito does not always transmit the disease. This lack of complete reproducibility suggests that there are other factors involved; and indeed, as we have seen, the known significant causes of malaria are quite complex, involving, as they do, factors of blood chemistry, general health, etc., in a way that is at present only partially understood.

The test of reproducibility enables us to tell why we have *not yet* included all of the significant causes. But there exists no test which could prove that we have included all of those causes. For it is always possible that the significant causes may include additional factors, as yet unknown, which have never yet varied sufficiently in the course of experiment and observations thus far carried out to change the effects appreciably. For example, in the nineteenth century it was thought that a person would have an adequate diet if he obtained a certain minimum quantity of fats, proteins, carbo-hydrates, and various minerals; and such a hypothesis was apparently verified by the fact that people obtaining an adequate supply of these materials from common foods suffered no visible nutritional deficiencies. But in a wider group of observations, in which it was noted, for example, that people who ate mainly rice from which the husks of the grain had been removed, suffered from the disease beri-beri, while people who ate the whole grain did not. It was therefore suspected that the husks of the grain contained additional substances needed in a complete diet. Later investigations disclosed the existence of a whole host of such substances, now called vitamins. The vitamins had indeed always been necessary for a healthful diet; but in most places they were so widely distributed that vitamin deficiencies had not been common enough to call attention to the existence of these very important needs of the human body. Thus, as the range of variation of experimental or observation conditions is widened, we must always be prepared for the possibility of dis-covering new significant causes in any particular field.

In order to deal with the problems raised by our inability to know all of the significant causal factors that may contribute to a given effect, there has evolved a distinction between immediate causes and conditions (or background causes). The immediate causes may be defined as those which, when subjected to the changes that take place in a given context, will produce a significant change in the effects. The conditions may be defined as those factors which are necessary for the production of the results in question, but which do not change sufficiently in the context of interest to produce an appreciable change in the effects. For example, one might say that fertile soil plus plenty of rainfall provides the general conditions (or background) needed for the growth of good crops. But the immediate cause would be the planting of the appropriate seeds.

The distinction between immediate causes and conditions is, however, an abstraction, useful for analysis but not strictly correct. For the background can always be changed, provided that conditions are altered sufficiently. We have seen, for example, in the case of the investigation of the cause of beri-beri, the origin of this disease had been confused by the existence of a general background in which most foods had enough vitamins for an adequate diet. But later investigations disclosed conditions in which this background did not exist.

Not only can background conditions be changed by external factors, but very often they can be changed significantly, after enough time, by the processes taking place in the background itself. For example, the cutting down of forests followed by the planting of crops may exhaust the fertility of the soil, and may even change the climate and the annual rainfall appreciably. In physics, the influence of any process on its "background" is even more strikingly brought out by Newton's Law that action and reaction are equal. From this law, it follows that it is impossible for any one body to affect another without itself being affected in some measure. Thus, in reality, no perfectly constant background can exist. Nevertheless, in any given problem a large number of factors may remain constant enough to permit them to be regarded, to an adequate degree of approximation, as forming a constant background. Thus, the distinction between immediate causes and conditions, or background causes, is relative and dependent on the conditions. Yet, because we can never be sure that we have included *all* of the significant causes in our theory, all causal laws must always be completed by specifying the conditions or background in which we have found that they are applicable.

5. MORE GENERAL CRITERIA FOR CAUSAL RELATIONSHIPS

Even when reproducible and controlled experiments are not possible, and even when the conditions of the problem cannot be defined with precision, it is still often possible to find at least some (and in principle an arbitrarily large number) of the significant causes of a given set of phenomena. This can be done by trying to find out what past processes could have been responsible for the observed relationships that now exist among these phenomena.

A very well-known example of a science in which reproducible and controlled experiments are impossible (at least with methods available at present), and in which the conditions of the problem cannot be defined very well, is geology. In this science, the most important method of formulating theories is to try to reconstruct the past history of the earth on the basis of observations of existing structures of rocks, mountains, seas, etc. We then ask, "What could have caused these present structures to be what they are?" We may see, for example, a set of layers of rock folded diagonally. The existence of such a structure suggests that the layers were deposited horizontally, when the region was at the bottom of a sea or a lake. The layers were then pushed up and folded over by the movements of the earth.

Although this explanation seems very plausible there is clearly no way to prove it by controlled and reproducible experiments or observations carried out under prescribed conditions, as all of the processes in question happened a long time ago, and the scale of the phenomena is, in any case, too large for us to do an experiment to verify such a theory. Moreover, because the number of geological formations available for study is limited, and because each formation has so many individual peculiarities that it is, to some extent, a problem in itself, we cannot hope that there would be enough naturally occurring variations in the various significant causes to substitute for an experiment with controlled variations under prescribed conditions.

Does this mean that there is no way to verify hypotheses concerning the causes of geological formations? Clearly not. First of all, there is the general consistency with which a very wide body of data can be explained. For example, the same type of assumption that would explain the folded structures of rocks in some places could also explain the fact that the shells of marine animals are often found at high altitudes, indicating that these regions were once below the sea, and further verifying the idea that over long periods of time the earth moves a great deal. Examples of this kind can be multiplied. Thus we obtain support for the theories of geology. Still more support can be obtained if the theories will correctly predict new discoveries. For example, according to certain theories of how oil was formed, we expect to find oil in certain types of places and not in others. If oil is fairly consistently found where predicted, and if it is not found where the theory says it should not be found, then we obtain an important verification of the hypotheses concerning the origin of oil.

Of course, hypotheses of the type that we have discussed above will, in general, be subject to corrections, modifications and extensions, which may have to be made later when new data become available. In this respect, however, the situation in geology is not basically different from that in fields where reproducible experiments and observations can be done under specified conditions. In such fields, too, hypotheses are subject to later corrections, modifications, and extensions. For example, even Newton's laws of motion,* which for over two hundred years were regarded as absolutely correct expressions of the most fundamental and universal laws of physics, and which had behind them the support of an enormous number of reproducible and very precise experiments and observations carried out under well-defined conditions, were ultimately found to be only an approximation. This approximation is very good at velocities that are low compared with that of light, but at higher velocities it ceases to be good. Here, one must use Einstein's theory of relativity, which yields approximately the same results as do Newton's laws of motion at velocities low compared with that of light, but which leads to completely different results at higher velocities. It goes without saying, of course, that in the future we may discover new conditions (not necessarily related to the velocity) in which the theory of relativity is found to be an approximation, which therefore has to be corrected, modified, and extended. Indeed, as was pointed out in Section 2, this is the normal pattern by which a science progresses, both on its theoretical and on its practical and experimental sides; i.e. by a continual application of the theory to new problems and new conditions, and by a continual revision and improvement of the theory in the light of what has been learned in these new applications.

In the last analysis, then, the problem of finding the causal laws that apply in a given field reduces to finding an answer to the question, "Where do the relationships among the phenomena that we are studying come from?" If reproducible controlled experiments or observations carried out under specified conditions are possible, these make available an important and very effective tool for verifying our hypotheses concerning the causal relationships. Whether such experiments are available or not, hypotheses can always be verified by seeing the extent to which they explain correctly the relevant facts that are known in the field in question, and the extent to which they permit correct predictions when the theory is applied to new phenomena. And as long as these possibilities exist, progress can always be made in any science towards obtaining a progressively better understanding of the causal laws that apply in the field under investigation in the science in question.

6. CAUSAL LAWS AND THE PROPERTIES OF THINGS

Thus far, we have been tending to centre our attention on the aspect

* We shall discuss these laws in more detail in Chapter II.

of the prediction of the *course of events* by means of causal laws; for example, the appearance of disease upon exposure to germs, the growth of seeds in proper soils, the improvement of health with changes in nutrition, the development of geological formations, etc. We shall now consider another equally important and indeed very closely related side of causality, namely, the predictions of the properties of things, both qualitative and quantitative.

Elementary aspects of this side of causality are met quite frequently in common life. Thus, an egg left in boiling water for a while will get hard; a hard brittle piece of glass heated to a high temperature becomes soft and malleable. Water cooled below a certain temperature becomes a solid, and heated above a certain temperature becomes a vapour. At a less elementary level, we have the chemical reaction of various substances to yield qualitatively new types of substances. We have also the hardening of metals by alloying or by heat treatment. There is no limit to the number of examples of this kind that can be found, but in all of them the essential point is that causal connections exist which permit the prediction of the new properties that things develop after they have undergone certain processes, treatments, reactions, etc.

The cases cited above all have in common that the new properties are predicted on the basis of the notion implicit in the concept of causality, that changes that have been found to take place in the past will occur again in the future if similar conditions are reproduced. Hence, while it is predicted that certain changes of properties will take place under certain conditions, the new properties themselves are not predicted; they are simply taken from the results of previous observations or experiments. A more subtle type of causal law is one that permits the prediction of some of the new properties of things even before these things have yet been observed or produced experimentally. For example, chemists studying a series of compounds of a certain type may notice a systematic variation in properties as one goes from one member of the series to the next. Thus, in the case of a certain class of hydrocarbons, the boiling-point decreases systematically as the number of carbon atoms in a molecule increases. It then becomes possible to predict that a new type of molecule having more carbon atoms than any of those yet produced will very probably have a still lower boiling-point. In physics, similar predictions can be made. Thus, it was discovered that there exist isotopes of each element, which are different kinds of atoms having the same chemical properties but different atomic weights. With the aid of physical theories concerning the motions of atoms, it was shown that different isotopes should diffuse at different rates when subjected to differences in concentration. On the basis of this predicted difference of properties of different isotopes, a method was then developed which made possible the large-scale separation of the two isotopes of uranium. This method of separation is one of the essential factors that makes a nuclear reactor possible. In connection with the same general subject, it was predicted on the basis of existing theory that uranium exposed to neutrons should be transformed into a new element, plutonium, that had not previously been observed or produced anywhere else. Many physical and chemical properties of this new element were predicted approximately. Examples of predictions of this kind are becoming more and more common all the time in many branches of physics.

The fact that such predictions are possible shows that the causal laws are not like externally imposed legal restrictions that, so to speak, merely limit the course of events to certain prescribed paths, but that, rather, they are inherent and essential aspects of these things. Thus, the qualitative causal relationship that water becomes ice when cooled and steam when heated is a basic part of the essential properties of the liquid, without which it could not be water. Similarly, the chemical law that hydrogen and oxygen combine to form water is a basic property of the gases hydrogen and oxygen, without which they could not be hydrogen and oxygen (just as water could not be water if it did not become hydrogen and oxygen when subjected to electrolysis). Similarly, the various quantitative laws are also an essential part of the things to which they appertain. Thus, some of the properties by which we recognize a liquid are the value of the temperature at which it boils, the value of its electrical conductivity, the value of its density, the values of the frequencies of the spectral lines that it absorbs or emits (which determine its colour), and by a great many other such quantitative properties. Likewise, the general mathematical laws of motion satisfied by bodies moving through empty space (or under any other conditions) are essential properties of such bodies, without which they could not even be bodies as we have known them. Examples of this kind could be multiplied without limit. They all serve to show that the causal laws satisfied by a thing, either when left to itself or when subjected to specified external conditions, are inextricably bound up with the basic properties of the thing which helps to define what it is. Indeed, we cannot conceive how a thing could even have any properties at all if it did not satisfy some kind of causal laws; for the mere statement that a thing has a certain property (for example, that it is red) implies that it will react in a certain way when it is subjected to specified conditions (e.g. the red object exposed to white light will reflect mostly red light). In other words, the causal laws that a thing satisfies constitute a fundamental and inseparable aspect of its *mode of being*.*

In order to understand just why and how the causal laws are so closely bound up with the definition of what things are, we must consider the processes in which things *have become* what they are, starting out from what they once *were* and in which they continue to change and to become something else again in the future. Generally speaking, such processes are studied in detail in a particular science only after it has reached a fairly advanced stage of development, while in the earlier stages the basic qualities and properties that define the modes of being of the things treated in that science are usually simply assumed without further analysis. Thus, in the earlier stages of the development of biology, the various classifications of living beings according to their basic properties and modes of life were simply accepted as eternal and inevitable categories, the reasons for the existence of which did not have to be studied any further. Later, however, there developed the theory of evolution, which explained many of the fundamental traits that define the mode of being of each species in terms of the process of transformation limited by "natural selection", a process in which each species has come to obtain its present character and which is presumably continuing, so that new species may appear in the future. Likewise in physics, the earliest steps involved the simple acceptance of certain characteristic properties of matter (e.g. density, pressure, electrical resistance, etc.), without further analysis, while later there came theories which explained and predicted these properties approximately in terms of processes taking place at the atomic level and at other deeper levels. As examples we may consider the prediction of the different rates of diffusion of different isotopes and the prediction of the properties of the new element, plutonium, both of which have already been cited in this section. Until recently, in physics, such explanations of properties and qualities have tended to be mainly in terms of inner processes of the types described above, i.e. processes which take place within matter, at deeper levels. However, lately there has developed a tendency to introduce evolutionary theories into physics, especially in connection with the efforts in the science of cosmology to explain how the particular segment of the universe that is at present accessible to our observations came to have its particular properties. These theories aim at the explanation of the formation of galaxies, stars, and planets, the explanation of the distribution of chemical elements in various parts

---

* Or, as we pointed out in Section 1, the inner character of a thing and its relationships to external causal factors are united in the sense that the two together are what define the causal laws satisfied by that thing.

of space, etc., in terms of an historical and evolutionary process, in which matter starting out in an earlier state gives rise to the cosmological order that we are now studying. Vice versa, in biology there has developed a growing tendency to explain various specific properties of living being in terms of processes (chemical, physical, etc.) taking place within the living organism. Similar trends are to be found in other sciences, such as chemistry, geology, etc. Thus, with the further development of the various sciences, we are obtaining a progressively better understanding of how the causal laws governing the various processes that take place in nature become indissolubly linked with the characteristic properties of things, which help define their modes of being.

7. ONE-TO-MANY AND MANY-TO-ONE CAUSAL RELATIONSHIPS

It is now necessary to consider more general types of causal relationships that do not determine the effect uniquely. In real problems, it is very rarely possible to deal with *all* the causes that are significant, even in a well-defined context, in which conditions (or the background) do not change appreciably. Usually we are able to treat only *some* of the significant causes. Naturally, as we have seen in Section 3, the effects are not completely reproducible and therefore not completely predictable. Nevertheless, just because we do not have at our disposal all the significant causes in a given problem, it does not mean that no predictions at all can be made. For, in such cases, it is generally possible to predict effects approximately, in the sense that they will be within a certain possible *range*. For example, if a gun is aimed at a certain point, the projectile does not land precisely at the place predicted by Newton's laws of motion (which are the causal laws that are pertinent in this problem). It is found, however, in a long series of similar shots, that the results cluster in a small region near the point that was calculated. A similar pattern of behaviour is demonstrated very generally in all fields in which causal laws are used for making predictions. For in every such prediction there is always a certain range of *error*, which may vary in a way that depends on the conditions of the problem, but which can never be eliminated completely. Thus, it is a general feature of causal relationships that they do not in reality determine future effects uniquely. Rather, they make possible only a one-to-many correspondence between cause and effect, in the sense that a specification of certain causes will in general limit the effect to a certain range of possibilities.

Of course, the fact that a causal relationship fails to determine future effects uniquely does not mean that nothing determines these effects. Indeed, this would be contrary to the principle that everything comes from other things (described in Section 1). In fact the more detailed determination of the effect depends on causes that lie outside the context of those that have been taken into account in the problem under investigation. In some cases, these additional causes could be taken into account with the aid of a more precise measurement of the causal factors already considered. Thus, in the case of aiming the gun, the first step in improving the precision would be to try to determine the angle of firing and the position of the gun more carefully. More generally, however, the precise determination of the effect eventually requires qualitatively new types of causal factors to be taken into account. For example, if we tried to obtain *unlimited* precision in the prediction of the trajectory of the shell, we should discover more and more significant factors on which this trajectory depended; e.g. the irregularities in the structure of the gun, air currents, small variations in temperature, pressure, humidity, and eventually even the motions of the molecules of which the gun, shell, air, and target are composed. Similar problems would arise in the effort to decrease the error in any causal prediction, with the purpose of obtaining unlimited precision. In other words, as we try to narrow down the range of a one-to-many causal relationship, we generally discover that each new order of magnitude of precision requires us to take into account new and qualitatively different causal factors on which the result depends.

In this connection, let us note that the one-to-many character of a causal law has no essential relationship to a *lack of knowledge* on our part concerning the additional causal factors to which the more precise details of the effect can be traced. Indeed, even if we did trace these details to such factors, so that we could make better predictions, it would still remain true that in the context in which these details do not appear, the law would continue to be valid in an objective sense as a one-to-many law. In other words, a one-to-many law represents an objectively necessary causal connection, but in this case, what is necessary is that *the effect remain within certain bounds*; and not, as in simpler types of causal laws, that the effect be determined uniquely.

Closely related to the one-to-many causal relationships are another type, which we may call the many-to-one causal relationships. A many-to-one causal relationship is one in which many different kinds of causes can produce essentially the same effect. An example is that all the rain that falls within a certain watershed will, independently of precisely where it drops, reach the sea in a certain place (i.e. where the main river of the watershed flows into the sea).

Likewise, independently of an enormous number of possible variations in the details of the environment in which a given creature lives, it can be predicted that this creature must eventually die. Examples of this kind are to be found in every field. Thus, in physics, if a body is disturbed or set into motion when it is near a position of stable equilibrium, it will eventually (because of friction) come back to its equilibrium position, independently of a wide range of possible initial motions, Indeed, in every field, all qualitative causal laws have a many-to-one character. For the prediction of a given quality may in general be made independently of a wide range of details, especially those of a quantitative nature. Thus, in the example of the transformation of water into steam, this transformation takes place independently of the quantity of heat supplied, provided that this quantity is more than that needed to furnish the so-called latent heat of evaporation (plus, of course, that needed to heat the water to the boiling-point). Moreover, not only qualitative but also quantitative laws may have a many-to-one character. Thus, the laws of thermodynamics deal with the properties of matter in thermal equilibrium. Quantitative relationships that are independent of the details of the processes by which equilibrium was attained are valid for equilibrium conditions.*

It must be remembered, however, that only some of the properties of an effect are unaffected by a wide range of variations in the causes. Indeed, according to the principle enunciated at the beginning of Section 1, no aspect of anything ever disappears completely without having *some* effect, so that it would be impossible for the two different causes to lead to completely identical results. Thus, if the water falling inside a particular watershed is stored in a dam, it might generate power, while if it is allowed to flow in its natural irregular path, it might instead flood the land and destroy cities. But independently of these details, the water in it will eventually reach the sea at the mouth of the main river in the watershed. Similarly, the ways in which a given creature lives will have effects on future generations as well as on the environment in general, even though, no matter what it does, it will die. Thus, while it is possible for certain aspects of an effect to come about independently of a wide range of causes, one discovers that as the effect is considered

---

* Many-to-one and one-to-many laws are interwoven into a unity, as they must be, because they both describe the same process. Thus, the laws of thermodynamics not only have a many-to-one character, but also a one-to-many character, coming from the possibility of error, originating in the fact that the cancellation of statistical fluctuations in the above motions (see Chapter II, Section 14) that give rise to the laws of thermodynamics, is never perfect. Similar interweaving is found on closer analysis in all cases of one-to-many and many-to-one laws.

either in more detail or in a broader context, each different kind of cause produces some difference in the effect.

The existence of one-to-many and many-to-one causal relationships is a very important characteristic of causal laws in general. To see one reason why this characteristic is so important, let us recall that incomplete precision in causal predictions comes from the fact that a given result depends on a great many factors that lie outside the context treated in a given problem. From a purely logical point of view, it would always be conceivable that these unknown, or at best poorly known, factors could produce variations in the effects of interest that went beyond any specified limits. Because, in such a wide range of fields, these factors do produce effects that stay within bounds, and which thus give rise to the one-to-many causal relationships, *it is possible to study a given problem, in some degree of approximation, without first taking into account the infinity of factors that are needed for a perfectly precise prediction of any given result.* The existence of many-to-one causal relationships evidently also contributes towards this possibility; since this means that many results can be studied independently of a very wide range of complicated details unknown to us or for other reasons too difficult to be studied under present conditions. We see, then, that the objectively one-to-many and many-to-one character of the causal relationships help to make it possible for us to have approximate knowledge about certain limited aspects of the world, without our first having to know everything about everything in the whole universe. And thus these causal relationships also help to make possible the characteristic scientific procedure of studying a problem step by step, each step laying the foundation for making the deeper, more detailed, or more extensive study that leads to the next.*

Within the general framework of one-to-many and many-to-one causal relationships, the one-to-one relationship is then an idealization which is never realized perfectly. Under certain limited conditions it may be approached so closely that, as far as what is

---

* A well-known example of this procedure occurs in physics. Thus, the first laws of physics to be discovered were those of macroscopic physics. Then, with the aid of these laws, the next step was to the laws of atomic physics. As we shall show in more detail in Chapter II, Section 10, the possibility of studying the laws of macroscopic physics without first knowing those of atomic physics comes from the many-to-one character of the statistical aspects of the laws of atomic physics, which permits a certain approximate autonomy of the laws of the higher level. The next step was to go, in a similar way, from the atomic level to the level of the nucleus, and now we shall see in later chapters (especially IV and V), physics seems ready to penetrate once again in a similar way to a still deeper level.

essential in the context of interest is concerned, we may consider the causal relationship as being approximately one-to-one. The nearest case known to a set of one-to-one causal relationships arises in connection with an isolated mechanical system, which can be treated in terms of Newton's laws of motion. These laws give a one-to-one connection between the positions and velocities of all the parts of the system at a given instant of time and their positions and velocities at any other instant of time.* This one-to-one connection is an idealization for several reasons. First of all, no mechanical system is ever completely isolated. Disturbances arising outside the system will destroy the perfect one-to-one character of the connection. Secondly, even if we could isolate the system completely, there would still exist disturbances coming from motions at the molecular level. Of course, one could in principle try to take these into account by applying the laws of motion to the molecules themselves, but then one would discover still further disturbances coming from the quantum-mechanical and other deeper-lying properties of matter.† Thus, there is no real case known of a set of *perfect* one-to-one causal relationships that could in principle make possible predictions of *unlimited* precision, without the need to take into account qualitatively new sets of causal factors existing outside the system of interest or at other levels.‡

8. CONTINGENCY, CHANCE, AND STATISTICAL LAW

Now contingencies are, as we have pointed out in Section 1, possibilities existing outside the context under discussion. The essential characteristic of contingencies is that their nature cannot be defined or inferred solely in terms of the properties of things within the context in question. In other words, they have a certain relative *independence* of what is inside this context. However, as we have seen, our general experience shows that all things are interconnected in some way and to some degree. Hence we never expect to find complete independence. But to the extent that the interconnection is negligible, we may abstract out from the real process and its interconnections the notion of *chance contingencies*, which are idealized, as completely independent of the context under discussion. Thus, like the notion of necessary causal connections, the notion of chance contingencies is seen to be an approximation, which gives a partial treatment of certain aspects of the real process, but which eventually has to be corrected and completed by a consideration of the causal

---

* These laws will be discussed in more detail in Chapter II.

† These will be discussed in Chapters III, IV, and V.

‡ In Chapter V we shall discuss the question of whether such relationships are in principle even possible.

interconnections that always exist between the processes taking place in different contexts.

In order to bring out in more detail what is meant by chance, we may consider a typical chance event; namely, an automobile accident. Now it is evident that just where, when, and how a particular accident takes place depends on an enormous number of factors, a slight change of any one of which could greatly change the character of the accident or even avoid it altogether. For example, in a collision of two cars, if one of the motorists had started out ten seconds earlier or ten seconds later, or if he had stopped to buy cigarettes, or slowed down to avoid a cat that happened to cross the road, or for any one of an unlimited number of similar reasons, this particular accident would not even have happened; while even a slightly different turn of the steering wheel might either have prevented the accident altogether or might have changed its character completely, either for the better or for the worse. We see, then, that relative to a context in which we consider, for example, the actions and precautions that can be taken by a particular motorist, each accident has an aspect that is fortuitous. By this we mean that what happens is contingent on what are, to a high degree of approximation, independent factors, existing outside the context in question, which have no essential relationship to the characteristic traits that define just what sort of a person this motorist is and how he will behave in a given situation. For this reason, we say that relative to such a context a particular collision is not a necessary or inevitable development, but rather that it is an accident and comes about by chance, from which it also follows that, within this context, the question of just where, when, and how such a collision will take place, as well as that of whether it will take place or not, is unpredictable.

So much for an individual accident. Let us now consider a series of similar accidents. First of all, we note that there is an irregular and unpredictable variation or fluctuation in the precise details of the various accidents (e.g. precisely when and where they take place, precisely what is destroyed, etc.). The origin of this variation is easily understood, since a great many of the independent factors on which the details of the accidents depend fluctuate in a way having no systematic relationship to what a particular motorist may be doing.

As the number of accidents under consideration becomes larger and larger, however, new properties begin to appear; for one finds that individual variations tend to cancel out, and statistical regularities begin to show themselves. Thus, the total number of accidents in a particular region generally does not change very much from year to year, and the changes that do take place often show a regular trend. Moreover, this trend can be altered in a systematic way by the alteration of specific factors on which accidents depend. Thus, when laws are passed punishing careless driving and requir-- ing regular inspection of mechanical parts, tyres, etc., the mean rate of accidents in any given region has been almost always found to undergo a definite trend downward. In the case of an individual motorist taking a particular trip, no very definite predictions can in general be made concerning the effects of such measures, since there are still an enormous number of sources of accidents that have not yet been eliminated; yet statistically, as we have seen, variations in a particular cause produce a regular and predictable trend in the effect.

The behaviour described above is found in a very wide range of fields, including social, economic, medical, and scientific statistics and many other applications.* In all these fields, there is a character- istic irregular fluctuation or variation in the behaviour of individual objects, events, and phenomena, the details of which are not pre- dictable within the context under discussion. This is combined with regular trends in the behaviour of a long series or large aggregate of such objects, events, or phenomena. These regular trends lead to what we may call *statistical laws*, which permit the approximate pre- diction of the properties of the "long run" or average behaviour of a long series or large aggregate of individuals, without the need to go to a broader context in which we would take into account additional causal factors that contribute to governing the details of the fluctuations of the individual members of such series of aggre- gates.

The tendency for contingencies lying outside a given context to fluctuate approximately independently of happenings inside that context has demonstrated itself to be so widespread that one may enunciate it as a principle; namely, the principle of randomness. By randomness we mean just that this independence leads to fluctuation of these contingencies in a very complicated way over a wide range of possibilities, but in such a manner that statistical averages have a regular and approximately predictable behaviour.†

It is clear, then, that when we know that a certain fluctuation is due to chance contingencies lying outside the context of the causal laws under discussion, we know more than the mere fact that the causal laws in question do not give perfectly accurate predictions; we know also that the contingencies will produce complicated

---

* We shall discuss some of these further applications in more detail in Chapter II, Section 14.

† For a more precise definition of randomness, see D. Bohm and W. Schutzer, *Supplemento al Nuovo Cimento*, Series X, n. 4, p. 1004 (1955).

fluctuations having regular statistical trends. Consider, for example, the problem of error in measurement discussed in the previous section. Such errors are generally divided into two classes, *systematic* and *random*. Systematic errors arise, but they are just due to external causes, and not to real chance contingencies that fluctuate independently of the context in question. To reduce systematic errors, we must obtain an improved understanding and control of the factors that are responsible for the error. The random part of the error can, however, be reduced simply by taking the average of more and more measurements. For, according to a well-known theorem, the effects of chance fluctuations tend to cancel out in such a way that this part of the error is inversely proportional to the square root of the number of measurements. This shows how the fact that a certain effect comes from chance contingencies implies more than the fact that the causes lie outside the context under discussion. It implies, in addition, a certain objective characteristic of randomness in the factors in which the effect originates.

We see, then, that it is appropriate to speak about objectively valid laws of chance, which tell us about a side of nature that is not treated completely by the causal laws alone. Indeed, the laws of chance are just as necessary as the causal laws themselves.* For example, the random character of chance fluctuations is, in a wide variety of situations, made inevitable by the extremely complex and manifold character of the external contingencies on which the fluctuations depend. (Thus random errors in measurement arise, as we have seen, in a practically unlimited number of different kinds of factors that are essentially independent of the quantity that is being measured.) Moreover, this random character of the fluctuations is quite often an inherent and indispensable part of the normal functioning of many kinds of things, and of their modes of being. Thus, it would be impossible for a modern city to continue to exist in its normal condition unless there were a tendency towards the cancellation of chance fluctuations in traffic, in the demand for various kinds of food, clothing, etc., in the times at which various individuals get sick or die, etc. In all kinds of fields we find a similar dependence on the characteristic effects of chance. Thus, when sand and cement are mixed, one does not carefully distribute each individual grain of sand and cement so as to obtain a uniform mixture, but rather one stirs the sand and cement together and depends on chance to produce a uniform mixture. In Chapter II, Section 14, we shall consider more complex examples connected with the motions of atoms to produce, partly with the aid of the cancellation of chance fluctuations,

---

* Thus, necessity is not to be identified with causality, but is instead a wider category.

uniform and predictable properties at the macroscopic domain (e.g. pressure, temperature, etc.). Here we shall see that the mode of being of matter in the macroscopic domain depends on the cancellation of chance fluctuations arising in the microscopic domain.

Not only are the regular relationships which come out of the tendency towards cancellation in a large number of chance fluctuations important, but under certain conditions even the fact that the chance fluctuations cover a wide range of possibilities in a complicated way may be extremely important. For one of the most characteristic features of chance fluctuations is that *in a long enough time* or *in a large enough aggregate*, every possible combination of events or objects will eventually occur, even combinations which would at first sight seem very unlikely to be produced. In such a situation, those combinations which result in some irreversible change or in some qualitatively new line of development are particularly significant, because once they occur, then the chance process comes to an end, and the system is irrevocably launched on its new path. As a result, given enough "mixing" or "shuffling" of the type connected with chance fluctuations, we can in such situations predict the ultimate result, often with impressive certainty.

A very interesting example of the property of chance described above occurs in connection with a current theory of the origin of life, suggested by Opharin. This theory is based on the hypothesis that perhaps a billion years ago or more, the atmosphere of the earth contained a high concentration of hydrocarbons, ammonia, and various simple organic compounds that would result from the combinations of these substances. Under the action of ultraviolet light, high temperature, electrical discharges, and the catalytic action of various minerals, these compounds would have tended to associate and to form ever more complex molecules. As the seas and the atmosphere were stirred up by storms and in other ways, all sorts of chance combinations of these compounds would have been produced. Eventually, after enough hundreds of millions of years, it would have been possible for just those combinations to occur which corresponded to the simplest possible forms of living matter. This point would, however, have been marked by a qualitative change that did not reverse; for the living matter would begin to reproduce at the expense of the surrounding organic material (since this is one of the basic characteristics that distinguishes living from non-living organic matter). From here on, the process would have been removed from the domain of pure chance. Moreover, as conditions changed, the living matter would start to evolve in accordance with the laws of transformation that have already been studied in considerable detail in biology; and eventually it would give rise to the manifold forms of life that exist today.

We see, then, the important rôle of chance. For given enough time, it makes possible, and indeed even inevitable, all kinds of combinations of things. One of those combinations which set in motion irreversible processes or lines of development that remove the system from the influence of the chance fluctuations is then eventually certain to occur. Thus, one of the effects of chance is to help "stir things up" in such a way as to permit the initiation of qualitatively new lines of development.

9. THE THEORY OF PROBABILITY

Just as the causal laws came to be expressed more precisely with the aid of certain kinds of mathematical formalisms (for example, the differential calculus), a characteristic mathematical instrument, known as the theory of probability, evolved for the expression of the laws of chance. In this section, we shall sketch briefly how this form of mathematics arose and what it means.

Historically, the notion of probability was first given a precise form in connection with gambling games. A good example is furnished by the game of dice. If we follow the results of each individual throw of the dice, we discover that they fluctuate irregularly from one throw to the next, in the way that is characteristic of chance events, as described in the previous section. As a result, we cannot predict what will be obtained in any given throw, either on the basis of the results of earlier throws, or on the basis of anything else that can be specified within the context of the game. Despite the unpredictable variations in the results of individual throws described above, however, gamblers have developed the custom of betting on a given combination, and of giving certain odds that depend on the combination in question. Experience has demonstrated that corresponding to each possible combination, there seems to exist a set of appropriate "fair odds", such that if these odds are offered, then in the long run the gambler will neither win nor lose systematically.

The problem that was attacked by the earliest mathematicians to turn their attention to this subject* was to find a theoretical way of calculating what these "fair odds" should be. In the case of the throws of a die, for example, this problem was solved by supposing that all six faces of each die are "equally likely" in each throw. Thus, the probability that a given die will come out a five is 1/6, and since the dice are "independent", the probability that both will come out fives is the product of the separate probabilities that each

* Among the earliest mathematicians to work with the concept of probability were Pascal, Fermat, Bernoulli, and Laplace.

one individually will come out a five, which is $1/6 \times 1/6 = 1/36$. Hence, the "fair odds" in this case are 36 to 1.

Although the method of solution of the problem indicated above certainly worked in connection with games of chance, it involved the introduction of the rather vague notion of equal "likelihood" or "equiprobability" of the various possible results of a throw. This notion initially contained a mixture of two very different interpretations of probability, which we may call respectively, the "subjective" and the "objective". In the subsequent development of the subject, these two interpretations became distinct; and in order to permit a clearer presentation of the essential ideas, we shall give here only the more definite interpretations that developed later.

In the subjective interpretation of probability, it is supposed that probabilities represent, in some sense, an incomplete degree of knowledge or information concerning the events, objects, or conditions under discussion. Thus, in the case of the game of dice, we have no way of knowing with certainty before the dice are thrown what the results of each individual throw will be (since these results are determined by the initial positions and velocities of the various parts of the dice in each throw which are not accessible to us in practice). Hence, if the dice are, as far as we can tell, symmetrically constructed, we know of no reasons favouring the suggestion that we will obtain any one side instead of another, and we therefore assign equal probabilities to each side. In this point of view, then, probability is regarded as something that measures or reflects a degree of our information, so that it is an essentially subjective category, which would cease to be necessary or even to have meaning if we could obtain precise knowledge concerning the initial motions of the dice in each throw.

The above interpretation of probability as representing nothing more than our own mental reflexes under conditions in which we do not have complete knowledge is, however, not adequate to treat an essential aspect of the problem of what is meant by probability. For it gives us no idea at all of why probability can be used to make approximate predictions about the actual *relative frequency* with which a given face of the die will be obtained after a large number of throws. Thus, the mere fact that we do not know any reasons that would favour one face over another does not by itself necessarily imply approximately equal relative frequencies for all possible results. Indeed, from the fact that we do not know anything at all about the initial motions of the dice as they are thrown, we can conclude only that we do not know anything at all about what the final results will be, not only in each individual case, but *also in an arbitrarily long series of cases.* For precisely among the things that we do not know about these initial motions there could conceivably exist a hidden tendency in them to favour one result over another. Vice versa, even if we were somehow able to know the initial conditions beforehand for each individual throw, this would not change the fact that in a typical series these conditions are in the long run and on the average distributed in such a way as to lead to approximately equal relative frequencies for each face. As a result, the theory of probability would in such cases provide a good approximation to the relative frequencies that would be predicted with the aid of perfect knowledge of the initial conditions determining each individual event.

Evidently, then, the applicability of the theory of probability to scientific and other statistical problems has no essential relationship either to our knowledge or to our ignorance. Rather, it depends only on the objective existence of certain regularities that are characteristic of the systems and processes under discussion, regularities which imply that the long run or average behaviour in a large aggregate of objects or events is approximately independent of the precise details that determine exactly what will happen in each individual case.

On the basis of the above considerations, we are then led to interpret the probability of, for example, a given result in the game of dice as an objective property associated with the dice that are being used and with the process by which they are thrown, a property that can be defined independently of the question of whether or not we know enough to predict what will happen in each individual throw. The significance of this property is that in the long run, and on the average, the relative frequency with which a given result will be obtained will fluctuate near a value that tends to come closer and closer to its probability. This, then, is the conception of probability that is relevant in statistical problems that arise in scientific research and in other fields. Of course, the word "probability" as commonly used also has the subjective meaning of describing how likely we think a given inference or conclusion drawn on the basis of incomplete knowledge may be. This meaning has, however, no essential relationship to the procedure by which we use the theory of probability in science and in other fields to make approximate predictions concerning relative frequencies of the various combinations of objects and events that occur in statistical aggregates, without the need to take into account precisely what each member of the aggregate is doing.

In order to understand in more detail the origin of the long run or average regularities that underlie the applicability of the theory of probability in games of dice (and in other gambling games), it is necessary only to note that in such games all the conditions are available for the applicability of the concept of chance and that of statistical law arising out of the effects of chance, as discussed in the previous section. Thus, if the die is thrown from an appreciable height, there is time for it to turn one or more times before it lands. The face on which it lands will then be sensitive to the initial motions, so that small variations in these motions can change any one final result into any other. Moreover, the human body, on whose motions the initial conditions are contingent, is a very complex system, whose functioning depends on an enormous number of varied kinds of fluctuating factors. Thus, it is quite understandable that in a large number of throws the initial motions transmitted from the hand to the die fluctuate sufficiently to make the final results vary over the full range of possibilities that are open. And since the multitude of factors in the human body are essentially independent of the initial orientations of the die, it is hardly surprising that in the long run and on the average no particular face tends to be favoured in these fluctuations, so that the individual throws fluctuate at random while statistical regularities appear in the mean relative frequencies with which each face is obtained. Thus, we have just the kind of dependence of the results in question on randomly fluctuating and independent contingencies lying outside the context under discussion, which is, as we have seen in the previous section, characteristic of chance phenomena.

With the aid of the concept of probability, it has been possible to develop an extensive mathematical theory, which yields expressions for the probabilities of complex combinations of events in terms of those of simpler events. This theory has demonstrated its utility in the many fields where there exist objects or events depending on chance contingencies arising outside the context under discussion. In the applications of this theory it must be remembered, however, that, as pointed out in Section 1, the causal laws and the laws of chance *together* are what bring about the actual development of things, so that either of them alone is at best a partial and approximate representation of reality, which eventually has to be corrected with the aid of the other.

10. GENERAL CONSIDERATIONS ON THE LAWS OF NATURE

We shall now give a brief summary of the essential characteristics of the laws of nature, as they have demonstrated themselves in the various examples considered in this chapter; and with the aid of this summary we shall obtain a further insight into the general structure of these laws.

Considerations on the Laws of Nature

First of all, our basic starting-point in studying the laws of nature was to consider the processes by which any one thing comes from other things in the past and helps to give rise to still other things in the future. Now this process cannot be studied in its totality which is inexhaustible, both in its quantitative aspects and in the complexity of its details. However, it is a fact, verified by human experience transmitted through our general culture since even before the beginnings of civilization, as well as by the experience of many generations of scientists, that parts of the processes described above can be studied approximately, under specified conditions, and in limited contexts. This is possible because there is an objective but approximate autonomy in the behaviour of these various parts of the processes relative to any particular context.*

When we study any particular set of processes within one of its relatively autonomous contexts, we discover that certain relationships remain constant under a wide range of changes of the detailed behaviour of the things that enter into this context. Such constancy is interpreted not as a coincidence, but rather as an objective necessity inherent in the nature of the things we are studying. These necessary relationships are then manifestations of the *causal laws* applying in the context in question. These laws do not have to determine a given effect *uniquely*. Instead, they may (in the case of one-to-many relationships) determine only that the effect must remain within a certain range of possibilities.

On the other hand, actual experience shows that the necessity of causal relationships is always limited and conditioned by contingencies arising outside the context in which the laws in question operate. These contingencies satisfy certain characteristic laws of their own: viz. the laws of chance, an approximate mathematical expression of which is given by the theory of probability.

Of course, by broadening the context, we may see that what were chance contingencies in the narrower context present the aspect of being the results of necessary causal connections in the broader context. But, then, these necessary causal connections are subject to still newer contingencies, coming from still broader contexts. Thus, we never really can eliminate contingencies. Rather, the categories of necessary causal connection and chance contingencies are seen to represent two sides of all processes. To consider only one of these sides, then, always constitutes an approximation that cannot apply

* The reasons for this autonomy will be discussed in some detail in Chapter V. We have already given some of them in connection with the discussion of one-to-many and many-to-one laws in Section 7 of the present chapter.

without limit, but that must eventually be corrected and supplemented by taking into account the other side.

The two sides of natural processes appear also in connection with statistical laws. Viewed from the side of chance contingencies, a statistical law is a regularity arising from the cancellation of chance fluctuations in a large aggregate of objects or events. But we may adopt the opposite view by considering the totality of all objects or events in a statistical aggregate as a single entity. The statistical laws, then, are approximate causal laws that apply to this new kind of entity. Thus, we see again how the same phenomena may be viewed from either side, depending on the context under discussion.

Besides having the two-sided character of necessity and contingency, the laws of nature show a richness of structure of a much more general character. Thus, considering the causal laws abstracted from contingencies, we find first of all that one obtains level after level of approximation, each involving qualitatively different kinds of causal factors. Even choosing only the significant causes that vary appreciably in the conditions of interest, we are still left with the possibility of one-to-many, many-to-one and one-to-one laws. Among the one-to-many aspects of the causal laws, one must consider the fact that every law has associated with it a certain error. This error arises essentially not because of a lack of knowledge on our part, but rather because of the neglect of objective factors existing outside the context under investigation. Even if we knew of these factors and could take them into account by going to a broader context, this would not change the fact that there exists a law, applying in the narrower context, which contains an error that would show up when we compared the predictions of the laws of the narrower context with those of the broader context. Then we have the many-to-one laws (such as qualitative relationships, statistical relationships, laws of thermodynamics, etc.) which have a certain degree of validity that is objectively independent of a wide range of details. Finally, there are the one-to-one relationships, which are abstractions that apply approximately in many cases when the many-to-one and one-to-many character of the laws can be neglected. Of course, all of these kinds of laws are interwoven into a unified fabric of law, as they have to be, since, after all, they apply to different aspects of the same sets of processes.

Then, when we do not abstract from contingencies, we must consider the laws of chance, which reflect all the richness of structure of the causal laws, as they must, since they treat just the opposite side of the same processes. Indeed, the interconnections among the various possible kinds of law are manifold and complex, involving laws reflecting laws, laws within laws (i.e. ever higher levels of accuracy), and laws which contain other laws as limiting and special cases. (E.g. relativity contains Newtonian mechanics as a limiting case when the velocity is small compared with that of light.) Moreover, this whole structure is an objective and necessary consequence of the very character of these laws, and is not just a special consequence of our way of thinking about things.

We may compare the structure of the totality of natural law to an object with a very large number (in reality infinite) of sides, having facets within facets, facets reflecting facets, facets consisting of mosaics of facets, etc. To know what the object is, then, we must have a large number of different kinds of views and cross-sections. Each view or cross-section then contributes to our understanding of many aspects of the object. The relationships between the views are, however, equally important, for they serve to correct the errors which arise as a result of regarding one or a limited number of views as a complete representation of the whole object; and they also indicate qualitatively new properties not apparent in the separate views (as two plane views of a scene taken from different angles permit us to infer its three-dimensional character). We see, then, that while each view and cross-section may vary depending on our own relationship to the object, we can obtain a closer and closer approximation to a concept of the real nature of the object by considering more and more views and cross-sections and their relationships. This concept, then, becomes less and less dependent on our own relationship to the object as the number of views and cross-sections is increased.

To pursue our analogy further, we may say that with regard to the totality of natural laws we never have enough views and cross-sections to give us a complete understanding of this totality. But as science progresses, and new theories are developed, we obtain more and more views from different sides, views that are more comprehensive, views that are more detailed, etc. Each particular theory or explanation of a given set of phenomena will then have a limited domain of validity and will be adequate only in a limited context and under limited conditions. This means that any theory extrapolated to an arbitrary context and to arbitrary conditions will (like the partial views of our object) lead to erroneous predictions. *The finding of such errors is one of the most important means of making progress in science.* A new theory, to which the discovery of such errors will eventually give rise, does not, however, invalidate the older theories. Rather, by permitting the treatment of a broader domain of phenomena, it corrects the older theories in the domain in which they are inadequate and, in so doing, it helps define the conditions under which they are valid (e.g. as the theory of relativity corrected

31

Newton's laws of motion, and thus helped to define the conditions of validity of Newton's laws as those in which the velocity is small compared with that of light). Thus, we do not expect that any causal relationships will represent *absolute truths*; for to do this, they would have to apply *without approximation*, and *unconditionally*. Rather, then, we see that the mode of progress of science is, and has been, through a series of progressively more fundamental, more extensive, and more accurate conceptions of the laws of nature, each of which contributes to the definition of the conditions of validity of the older conceptions (just as broader and more detailed views of our object contribute to defining the limitations of any particular view or set of views).

At any particular stage in the development of science, our concepts concerning the causal relationships will then be true only relative to a certain approximation and to certain conditions. Indeed, it is for just this reason that so many different kinds of explanations and theories applying to the same set of phenomena are possible. Each different theory or explanation focuses on a certain aspect of the laws of nature that is important under certain conditions, and treats this aspect within a certain degree of approximation. But *to the extent that different theories and explanations treat the same domain and to the approximation with which they are able to do this, they must agree.* Hence, the possibility of so many different kinds of explanations and theories of the same sets of phenomena does not imply that the laws of nature are arbitrary or conventional rules, that could be changed at will in accordance with our tastes, or with what is convenient for us in various kinds of problems. Rather it is merely a consequence of the infinite richness of the real relationships existing in natural processes and of our need to express partial aspects of these infinitely rich relationships in terms of finite laws based on experiments and observations done up to a particular period of time, which can reflect adequately only a limited part of the infinite totality that exists in nature.

11. CONCLUSION

In conclusion, it should be clear that the existence of natural law as we have described it in this chapter is of greatest importance in all branches of science.

In addition, however, the possibility of a science also depends on the particular structure of the laws of nature (e.g. existence of many-to-one and one-to-many laws, as well as other characteristics to be described in Chapter VI) which is such that there exist relatively autonomous contexts that can be studied separately to some degree of approximation, without our first having to learn everything about everything with perfect precision. In particular, this characteristic of the causal laws is the objective factor determining the division of the study of the world among the various sciences, as well as the concepts and methods that are appropriate to each particular science. Then, in a given science, it is this property of the natural laws which makes possible the existence of various branches, domains, and levels, all having an approximate autonomy. However, since the natural laws imply some kind of interconnection of all aspects of the world, as well as their approximate autonomy, this means that wider studies carried out in broader domains or in wider contexts permit a demonstration of the relationships between the various branches, domains, and levels in a given science, and between the various different sciences, as well as a penetration to new domains not hitherto known or investigated.

CHAPTER TWO

Causality and Chance in Classical Physics :
The Philosophy of Mechanism

1. INTRODUCTION

THE previous chapter was devoted to a general discussion of causality and chance. We shall now proceed to show in some detail how these categories manifest themselves in classical physics (which is roughly that branch of physics that had its fundamental development between the sixteenth and nineteenth centuries, inclusive). This subject is not only of considerable interest in itself, but it also has a great deal of bearing on the questions concerning the applicability of the concept of causality that have been raised during the twentieth century in connection with the quantum theory. For, as we shall see in later chapters, the inadequacy in the microscopic domain of the mechanistic form of determinism into which causality was restricted by classical physicists helped to provoke a very strong reaction in the opposite direction, and thus helped to encourage modern physicists to go to the opposite extreme of denying causality altogether at the atomic level. It will therefore be worth our while here to study fairly carefully just what were the notions of causality and chance that came to be associated with classical physics and just what were the problems that arose in connection with the applications of these notions. Then, in later chapters, but especially in Chapter V, we shall criticize the mechanistic point of view and develop in some detail a more general point of view, in which the problems described above do not arise.

2. CLASSICAL MECHANICS

The most important developments in physics between the sixteenth and the nineteenth centuries were founded on a certain body of very general, very comprehensive, and very precisely expressed theory usually referred to as *classical mechanics*, which concerns itself primarily with the laws governing the motions of bodies through

34

space. Now in the earlier stages of their development (as in the times of the ancient Greeks), the laws of mechanics had generally been given a vague and qualitative form.* With the work of Galileo and others, however, the tendency to express the laws of physics (and to some extent of chemistry and other sciences) in a precise quantitative form began to assume a very great importance. This trend towards quantitative precision in the expression of physical law continued to grow, and first reached its full development with Newton's laws of motion. These laws, which state that the acceleration of a body is directly proportional to the force acting on it and inversely proportional to its mass, are expressed mathematically by means of the differential equation,

$$\vec{F} = m\frac{d^2\vec{x}}{dt^2}$$

where $\vec{x}$ is the position vector of the body, $\vec{F}$ the force acting on it, and m is its mass.

Newton's laws of motion imply that the future behaviour of a system of bodies is determined completely and precisely for all time in terms of the initial positions and velocities of all the bodies at a given instant of time, and of the forces acting on the bodies. These forces may be *external forces*, which arise outside the system under investigation, or they may be *internal forces* of interaction between the various bodies that make up the system in question.

In many problems the external forces are small enough to be neglected (i.e. the system may be regarded as isolated), while the internal forces can be represented solely in terms of functions of the positions and velocities of the centres of mass of the bodies. This approximation is particularly good in the problem of the motion of the planets around the sun. In such a case, Newton's laws determine the future motions of the bodies in terms of nothing more than the positions and velocities of the bodies at a given instant of time. Thus, they constitute a set of "one-to-one" causal relationships, of the type described in Chapter I, Section 7. For given the complete set of causes (i.e. the initial positions and velocities of each body), then the complete set of effects (i.e. the later positions and velocities of each body) is determined uniquely.

More generally, however, no system of bodies is ever really completely isolated, nor can the forces between bodies ever really be

---

* Some precise expression of the laws of statics had already been given, even by the ancient Greeks (e.g. by Archimedes), but the laws of mechanics in general were expressed quite vaguely, and in most cases incorrectly.

expressed in terms only of the positions and velocities of the centres of mass of these bodies. Thus, even in astronomy, the isolation of the solar system is not perfect. Distant stars have some effects, even though these are small, while comets from interstellar space may occasionally enter the solar system and even deflect the orbit of a planet appreciably. Similarly, there is a small tidal friction which makes the planetary motions depend slightly on the configurations of land, water, and other fluids on the planets, and which is slowly causing the planets to circle closer to the sun and the moon to the earth. In all other problems, one likewise finds that the system is never completely isolated, and that it can never really be analysed into bodies for which the external motions are completely independent of the internal motions.* Thus, generally speaking, the complete set of causes needed to determine the future motions uniquely must include both the initial positions and velocities of the bodies and the various forces, both external and internal, that act on the bodies.

3. THE PHILOSOPHY OF MECHANISM

It is clear that Newton's laws of motion represented an enormous progress in the expression of causal relationships in the science of mechanics. For in place of crude and qualitative laws of mechanics that had been characteristic of the ancient and medieval periods, Newton's laws evidently represent a fundamentally new kind of law, making possible precise quantitative predictions, which permit a much more accurate test of the law, and serve as correspondingly precise guides in our efforts to alter and control the behaviour of mechanical systems.

The very precision of Newton's laws led, however, to new problems of a philosophical order. For, as these laws were found to be verified in wider and wider domains, the idea tended to grow that they have a *universal* validity. Laplace, during the eighteenth century, was one of the first scientists to draw the full logical consequences of such an assumption. Laplace supposed that the *entire universe* consisted of nothing but bodies undergoing motions through space, motions which obeyed Newton's laws. While the forces acting between these bodies were not yet completely and accurately known in all cases, he also supposed that eventually these forces could be known with the aid of suitable experiments. This meant that once the positions and velocities of all the bodies were given at any instant of time, the future behaviour of everything in the whole universe would be determined for all time. Laplace then imagined a superior being who could know all these positions and velocities, and who

* In this connection, see Section 13 of this chapter.

could calculate with complete precision everything that would happen in the universe. Thus, for this being, nothing unexpected could ever come into the world, since everything, even in the infinite future, would happen in a way that had been predetermined and, indeed, predetermined throughout the entire infinite past.

Here we have an interesting and important new development. For as long as Newton's laws of motion are applied to some limited system or domain, they merely form the basis of the science of *mechanics* which expresses in a precise mathematical form the causal laws that apply to that particular system or domain. When expressed in this manner, the science of mechanics evidently does not necessarily imply a *completely* determinate prediction of the future behaviour *of the entire universe.* For besides the fact that we are treating only a specified mechanical system to what must in general be only a limited degree of approximation, we must also consider the possibility that in new domains of phenomena or under new conditions not yet studied in physics, it is possible that newer and more detailed expressions of the laws of nature may be needed, expressions which might not even be possible in terms of the general mathematical and physical scheme underlying Newton's laws of motion. Thus, the conclusion that there is absolutely nothing in the entire universe that does not fit into the general theoretical scheme associated with Newton's laws of motion evidently has implications not necessarily following from the science of mechanics itself, but rather from the *unlimited* extrapolation of this science to all possible sets of conditions and domains of phenomena. Such an extrapolation is evidently then not founded primarily on what is known scientifically. Instead, it is in a large measure a consequence of a *philosophical* point of view concerning the nature of the world, a point of view which has since that time come to be known as mechanism.

Now, as we shall see in this chapter and in other parts of the book, the mechanistic philosophy has taken many specific forms throughout the development of science. The most essential aspects of this philosophy seem to the author, however, to be its assumption that the great diversity of things that appear in all of our experience, every day as well as scientific, can all be reduced completely and perfectly to nothing more than consequences of the operation of an absolute and final set of purely quantitative laws determining the behaviour of a few kinds of basic entities or variables. In this connection it must be stressed, however, that the mere use of a purely quantitative theory does not by itself imply a mechanistic point of view, as long as one admits that such a theory may be incomplete. Hence, mechanism cannot be a characteristic of any theory, but rather, as we have already stated above, a philosophical attitude towards that theory. Thus, it would have no meaning to say, for example, that Newtonian mechanics is mechanistic; but it has meaning to say that a particular scientist (e.g. Laplace) has adopted a mechanistic attitude towards this theory.

The first known form of mechanism was the atomic philosophy of Democritus and Leucippus, in which it was assumed that everything in the universe could be reduced to nothing more than the effects of the motions of atoms through space. The idea underlying Laplacian determinism was essentially the same one, with the addition, however, of the assumption that motions of these atoms are governed by Newton's laws, so that a precise calculation of the future behaviour of the universe is in principle possible. In this point of view, then, all the various qualitative properties that appear at the large scale, such as hardness, fluidity, colour, texture, etc., are regarded as purely subjective categories, since they do not appear in the basic laws governing the motions of the atoms, but are thought of as nothing more than intermediary concepts that we find it convenient to utilize in our thinking about the arrangements and motions of molecules *en masse*. The fundamental properties that are objective and not the result of our special ways of thinking about things, are assumed to be the basic quantitatively specifiable properties of the atoms—their positions, velocities, sizes, shapes, masses, the laws of force between them, etc.

In this early form of the mechanistic philosophy, the basic elements out of which the world was assumed to be constructed were effectively conceived of as mechanical parts, each of which has its place in a universal machine (which is frictionless because the laws of classical mechanics imply the conservation of mechanical energy). The nature of these parts is rigidly fixed and does not grow out of the context in which they are placed, nor does it change as a result of the actions of other parts. In this sense, the universal frictionless mechanism is an idealization of the machines with which we are familiar, for the latter are not frictionless, nor are their various parts unaffected by the actions of other parts (e.g. they break, wear out, etc.).

Now, as we shall see in later chapters, the philosophy of mechanism eventually came to be a very serious restriction on the further development of science. Nevertheless, for its time, it was an enormous step forward from the scholastic form of the Aristotelian philosophy that was prevalent during the Middle Ages. For in the scholastic philosophy every different thing and every different property and quality of things was conceived of as separate and completely distinct from all the others. Thus, scientific investigations guided by this philosophical point of view tended to consist mainly of the arrangement of things into various systems of classifications, which were regarded as eternal and unvarying in their nature. The mechanistic philosophy, however, suggested that beneath all this diversity, disparity, and apparent arbitrariness of qualities existed a set of simple and rationally understandable universal mechanical processes. These processes explained why things took the diverse forms that they did, and why they underwent the transformations that they did, in a way that was in principle subjected to complete calculation and verification, and that in practice could at least be calculated and verified within some degree of approximation. Thus, the mechanistic philosophy made possible a much more *unified* and *dynamic* point of view towards the universe than had been available during the Middle Ages. This point of view enabled one to see clearly the close relationships existing between a wide range of problems not even considered in the scholastic philosophy (e.g. the precise predictions of the influence of one planet on the motions of others, the determination of the trajectories of shells fired from guns, etc.).

4. DEVELOPMENTS AWAY FROM MECHANISM IN CLASSICAL PHYSICS

Even during the period of the greatest triumphs of mechanism, physics began to develop in new directions, tending to lead away from the general conceptual framework that had been associated with the original form of the mechanistic philosophy. The most important of these developments were those connected with the formulation of the basic laws of the electromagnetic field, with the elaboration of the kinetic theory of gases, and with the initiation of the use of statistical explanations for the laws of thermodynamics and other macroscopic properties of matter, rather than completely determinate types of explanations that had previously been the ideal in physics.

Although none of these developments stood in complete contradiction with a mechanistic point of view, each of them showed the need for a progressive enrichment of the basic concepts and qualities which were needed for expressing the laws of physics as a whole. The need for such enrichment time after time could in principle already have suggested to physicists that their basic philosophical point of view was not really adequate for the understanding of nature as a whole. Of course, it did not actually do this, because physicists made various adjustments, compromises, and extensions of their concepts, each time supposing that at last they had reached the ultimate general conceptual framework and system of basic qualities and motions that would, once and for all, permit the expression of the absolute and final laws of physics. Thus, they were enabled to retain an essentially mechanistic point of view, in spite of the many changes that occurred in the basic formulation of physical laws.

Throughout the rest of this chapter, then, we shall give a fairly detailed account of these later developments, taking some pains to bring out clearly just how the mechanistic philosophy can accommodate itself to deal with them. Further criticism of the mechanistic philosophy will then be deferred until later chapters, especially Chapter V, where we shall propose an alternative point of view.

5. WAVE THEORY OF LIGHT

One of the first new developments in physics that helped lay the foundation for important steps away from mechanism was the wave theory of light. This development was important for two reasons: first because it helped give rise to field theories (to be discussed in

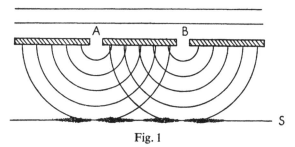

Fig. 1

Section 6), and secondly because it provided a set of concepts which were to prove to be of crucial importance in connection with the quantum theory (to be discussed in Chapter III).

Now, during the time of Newton it was not known whether light consisted of small particles moving very rapidly in straight lines (except when they were refracted or reflected by material bodies) or of a form of wave motion. Gradually, however, evidence accumulated suggesting that light is a form of wave motion. The most important of this evidence came from the experiments demonstrating the existence of interference. A parallel beam of light is allowed to fall on a slit, A (see Fig. 1). Some of the light passes through the slit and reaches a screen, S. Another slit, B, is then opened. Now, if light consisted of a rain of small particles, the region on the screen between slits A and B should everywhere be illuminated more intensely than if slit A alone were open. For to these particles of light reaching the screen from A must be added those coming from B.

On the other hand, in reality, one finds on the screen a set of alternate bright and dark fringes fairly close together. The fringes disappear when one of the slits is closed. This shows that when the light from A and from B come together, the net result may be an absence of light. Such a result would be very difficult to understand if light consisted simply of particles. However, it would be easy to understand if one assumed that light was a form of wave motion. For at certain points on the screen the waves coming from A could cancel those coming from B, producing darkness, while at other points they could add up, producing more light. The wave theory explained not only this experiment, but a great many others, in a quantitative way. It also made possible the calculation of the length of these waves, which was found to be of the order of 5×10^{-5} cms. The very short length of these waves explained why light usually seems to go in straight lines and to act as if it were made of particles. By analogy, one may consider waves in water. Very short water waves will be stopped by a barrier which is appreciably bigger than the length of the waves. But a wave that is much longer than the barrier will go around the barrier. Light shows a similar behaviour. Thus, if a very fine wire is viewed with the aid of a distant light the outlines of the wire seem to be indistinct. This phenomenon is known as diffraction. But if we take a large object, such as a house, the bending of the light waves as they pass the edge of the house is negligible and the light effectively goes in a practically straight line, as if it were made up of small particles moving in a straight line.

If light consists of waves, the question then arises: "How are these waves carried?" Since light is transmitted through a vacuum, it is clear that light-waves are not carried in any common material medium, as are the waves of water and sound. We shall consider this question further in Section 7, after we have discussed field theories in Section 6.

6. FIELD THEORY

We are now ready to consider the development of the first important new type of causal law that occurred during the nineteenth century; viz. that of field theory.

Recall that, in Newtonian mechanics, one always postulated that things were constituted of bodies interacting according to specified forces. A knowledge of the position and velocities of the bodies in a given isolated system would then permit us to predict all the motions taking place within that system. But throughout the nineteenth and early twentieth centuries, the need for considering new kinds of causal factors gradually came to be recognized. For in order to express the laws of electricity and magnetism, it was found that the

Newtonian scheme of bodies was not adequate. It was possible, however, to treat this problem by introducing, in addition to the bodies, a new set of entities known as the electric and magnetic fields. Whereas the mode of existence of the bodies required that they be localized in some definite region of space, the fields were conceived of *as continuously distributed* throughout space as a whole. At each point in space and at each instant of time, however, the components of the electric and magnetic fields were assumed to have definite values. The values of the components of the electric field at a given point were defined in terms of the force that would be exerted on a unit electric charge placed at the point in question, while the values of the components of the magnetic field were defined in terms of the force that would be exerted on a unit magnetic pole at that point.

Now as long as the fields are static, the electric and magnetic fields remain clearly separated and independent of each other. When the fields are changing with time, however, characteristic new phenomena appear. For example, if electric charges are in motion, we obtain an electric current and with it a *magnetic* field. Similarly a magnet in motion creates an electric field. Thus, electric and magnetic fields are not independent entities, but, rather, each helps to determine the other. Indeed, the experiments of Faraday disclosed a set' of precisely specifiable quantitative relations between the magnetic and electric fields. Maxwell, basing himself in part on these experiments, later extended the relations found by Faraday and developed a set of partial differential equations, now known as Maxwell's equations. These equations determine how the fields will change in terms of the values of the field quantities at each instant of time and in terms of the motions of all of the charged bodies in the system. But since the electric and magnetic fields contribute to the forces acting on the bodies, it is clear that fields and bodies co-determine each other. The combined laws (Newton's equations for the bodies plus Maxwell's equations for the fields) then form a unified and extended set of basic causal laws, generalizing the laws of Newton, which, as we recall, were expressed solely in terms of the motions of the bodies. Thus, the complete causal laws now include both bodies and fields.

The field theory of Maxwell led to many new predictions, that were later verified experimentally. One of the most important of these was that waves should be possible, in which the electric and magnetic fields oscillated in amplitude, in much the same way that the height of a pool of water oscillates when there is a wave on the surface of the water. It was deduced from Maxwell's equations that these waves should be propagated with a definite and predictable velocity, which could be calculated on the basis of numbers coming out of measurements that had already been made in connection with electrical and magnetic fields. When this number was calculated, it was found that the predicted velocity of these electromagnetic waves was, within experimental error, equal to the measured velocity of light. The fact that light is a wave has already been suggested by experiments on interference; but now the theory of Maxwell went further, for by predicting the velocity of light solely on the basis of measurements made in electricity and magnetism, it created strong evidence that waves of light were just the kind of waves predicted by Maxwell's equations.

Since the time of Maxwell, an enormous amount of experimental evidence corroborating the electromagnetic theory of light has been accumulated. These experiments, which cover a wide range of fields, extending from optics and infra-red spectroscopy through ultraviolet rays, X-rays, gamma rays, etc., provide a very convincing set of confirmations of the theory of Maxwell.

7. ON THE QUESTION OF WHAT IS THE NATURE OF THE ELECTROMAGNETIC FIELD

Having seen that there is a great deal of evidence suggesting that light consists of waves in the electromagnetic field, we are now led to raise the question: "What is the electromagnetic field?" Faraday, Maxwell, and other scientists of the time had postulated that all space is full of a very fine medium that they called the "ether". They supposed that, like air and water, this medium could support internal stresses and could therefore undergo wave motion. The internal stresses of the ether manifested themselves to us as electric and magnetic fields. They also hoped in this way to explain gravitational forces, as a different kind of internal stress of the ether.

Many experiments were therefore done with the purpose of finding direct experimental evidence of the existence of the ether, among the most famous of which was the Michelson-Morley experiment. We cannot go into details here, but shall merely state the net conclusion: viz. that all experiments aimed at detecting the ether gave negative results. Thus no experimental proof that the ether really exists could be found. This created a serious problem. For real physical effects such as light and gravitation could be transmitted for long distances through apparently empty space. To emphasize the seriousness of the problem, let us note that a steel cable having the diameter of the earth would not be strong enough to hold the earth in its orbit around the sun. Yet the gravitational force that holds the earth in its orbit is transmitted across 93,000,000 miles of space without any traces of a material medium in which these forces might be carried.

As yet, the problem of what material medium, if any, carries the electromagnetic field cannot be said to have been dealt with completely. What has happened is that scientists have gone around the problem. For, as time passed, it gradually became clear that to make theories of the ether without experimental clues as to what the ether might be was largely empty speculation. Instead, what was done was simply to assume the existence of the fields without reference to the question of whether or not the ether existed. The fields could in principle be defined at each point in space, and their variation in time was determined by Maxwell's equations. But as far as all physical phenomena that had yet been observed were concerned, the question of a material medium or "ether" in which these fields might be represented as states of stress or motion was irrelevant. Thus, even if an ether did exist, then at least within the context of the experiments that were possible at the time, all results would have been just the same as if it did not exist. In other words, all that had been significant thus far were the fields themselves. As a result, there arose the notion that the fields are qualitatively new kinds of entities, which we have the same right to postulate as we have to postulate material bodies (such as atoms), provided that such a postulate will help in the explanation of a large range of facts and experimental results. This point of view, which seems to have been suggested first by Lorentz, was later carried much further by Einstein. It is now held by a majority of physicists.

The introduction of the fields evidently involves a fundamental modification of our concept of matter and of space. Thus, the field concept implies that even when space contains no bodies as we know them, it could still be the site of continuously varying fields. These fields can be shown to carry energy, momentum, and angular momentum, so that they are even able to simulate some of the properties of moving bodies.* Indeed, Einstein has gone further; for he has made the very interesting suggestion that special kinds of fields† might exist, having modes of motion in which there would be pulse-like concentrations of fields, which would stick together stably, and would act almost exactly like small moving bodies. He further proposes that perhaps the so-called fundamental particles of physics, such as electrons and protons, may consist of such modes of motion

* Thus, the momentum of a beam of light leads to a measurable *radiation pressure* on a surface analogous to the pressure that would be exerted on this surface by a rain of molecules.

† These would be fields satisfying non-linear equations. See Chapter III, Section 3, for a discussion of some of the properties of non-linear equations.

of the fields. Whether we accept these proposals or not, however, it is clear that many of the basic properties of a material system, which determine its characteristic modes of behaviour (e.g. the forces acting on the various bodies within the system, the total energy, momentum, and angular momentum of the system, etc.) depend just as much on the fields as on the bodies. Thus, the concept of matter has effectively been expanded to include the notion of the field as representing the extension through a broad region of space of certain manifestations of a material system.

8. FIELD THEORIES AND MECHANISM

The hypothesis made towards the end of the nineteenth century that the field exists in its own right as a qualitatively new kind of entity was an important step away from mechanism. To be sure, as early as the eighteenth century, the concept of density and velocity fields had been used in hydrodynamics. But these fields were regarded as being nothing more than an approximate means of dealing with certain over-all properties of the molecules constituting the fluid. The same point of view was carried into electrodynamics by Faraday, Maxwell, and other scientists of the time when they supposed that the electromagnetic field represented nothing more than states of motion or of internal stress in the ether. The notion that the fields represented entities that had a certain existence in their own right, then constituted a genuine enrichment in the conceptual basis of physics. For in addition to formulating physical laws in terms of the motions of bodies through space, physicists now also formulated such laws in terms of a qualitatively new kind of motion, namely, that of involving a set of changing amplitudes of fields at various points in space.*

* It is true that it is not absolutely necessary to use the field concept in the problems described here. Thus, it is possible to eliminate the fields in terms of the motions of the particles, with the aid of the retarded potentials. This leads to an expression of the force acting on a given body at a certain time in terms of the motions of all the bodies, over a period of time that becomes infinite if we wish thus to take into account the effects of sources of electromagnetic radiations that are arbitrarily far away from the body in question.

To specify the motions of all bodies over all time is, however, not only a clumsy procedure, which would obviously be inadequate to treat even a simple problem such as the propagation of a radio wave along a wave guide; but it also does not seem to be the specification that corresponds to the form that physical laws take as one considers a broader or deeper range of problems. For example, black-body radiation strongly suggests that there is a field that absorbs the energy emitted by matter since the mean energy absorbed is exactly what would be absorbed by a set of equivalent oscillators. When we come to quantum electrodynamics, the field concept demonstrates important further advantages. For here the

On the other hand, as field theories came to be an accepted part of the structure of modern physics, a great many physicists began to give them what is, in its essence, a mechanistic interpretation. For instead of assuming that the whole of nature can be reduced to the motions of a few kinds of bodies, they assumed that the whole of nature can be reduced to nothing more than a few kinds of bodies and a few kinds of fields. Or with Einstein, they may have assumed that the whole of nature could be reduced to nothing more than fields alone. Thus, these physicists effectively argued that the philosophy of mechanism was right in general but wrong in the detail that it had previously left out an important set of mechanical parameters, the fields, which were actually needed for the complete specification of the state of everything in the whole universe.* It is true that the fields, being continuous, required a non-countable infinity of variables for their mathematical expression.† Thus, the mechanistic programme of predicting the future behaviour of the universe by knowing the initial values of all the mechanical parameters involved (in this case, those connected both with the fields and with the bodies) was now clearly impossible in practice. Nevertheless, this behaviour could still be conceived of as determined in principle by these mechanical parameters. Thus, one could also imagine that the super-being of Laplace was endowed with the power of dealing with a non-countable infinity of variables, so that he could then calculate the future of the universe with complete precision, although the labour involved would obviously be infinitely greater than a universe consisting of nothing but bodies. It

very existence of particles is understood in terms of the notion of quantized states of the fields, so that one can likewise understand the "creation" and "destruction" of such particles as changes in the state of excitation of the fields. Similarly, the quantum fluctuations of the vacuum, which have demonstrated their importance in many experiments, are likewise described in a very natural way in terms of the field concept. On the other hand, the "creation" and "destruction" of particles cannot even be treated in terms of retarded potentials while the treatment of the quantum fluctuations in these terms is very artificial. Thus, the field point of view is strongly favoured in a rather wide range of problems.

* This view is at present most frequently embodied in the formulation of the laws of nature in terms of a "variational principle", in which these laws are derived by minimizing a "Lagrangian". To treat the fields, one then simply adds the "field Lagrangian" to the "particle Lagrangian". Thus, the "field co-ordinates" are regarded as additional mechanical variables that have to be added to the particle variables to constitute a generalized mechanical system.

† If we consider a system enclosed in a box, it is true that the field variables become countable (e.g. a Fourier series). However, to treat the universe *as a whole*, we are not permitted to assume such a box. Hence, the variables are not countable in the problem under discussion here.

is clear, then, that the point of view described above retains the most essential and characteristic feature of mechanism (see Section 3), namely, to reduce everything in the whole universe completely and perfectly to purely quantitative changes in a few basic kinds of entities (in this case, bodies and fields, or fields alone, as in the point of view of Einstein), which themselves never change qualitatively. It is, to be sure, a more subtle and elaborate form of mechanism than that of Laplace, but in becoming more subtle and elaborate, it did not cease to be mechanistic.

9. MOLECULAR THEORY OF HEAT AND THE KINETIC THEORY OF GASES

Along with the field theory and the theory of light there developed another new branch of physics, namely the molecular theory of heat and the kinetic theory of gases, which, as we have already pointed out, also played an important rôle in the gradual process of undermining mechanism that took place during the eighteenth and nineteenth centuries.

As is well known, it was shown by Mayer, Joule, and others that water is heated when it is set in agitated and turbulent motion, for example, by a moving paddle-wheel. Joule then measured the heating which results when such a wheel is turned by a weight that is allowed to fall through a predetermined distance. The heat produced was found to be always proportional to the energy liberated by the falling weight. Vice versa, heat engines were constructed (e.g. steam engines) which turned heat into mechanical energy. Again there was the same proportionality between heat and mechanical energy.

A theory of heat was then developed. It was postulated that heat is a form of chaotic molecular motion. Thus, as the paddle-wheel turned, it created vortices in the water. These vortices gradually became smaller and more and more chaotic until they approached a molecular level in size. At this stage, the energy of the regular mechanical motion of paddle-wheel had been transformed completely into energy of irregular or chaotic molecular motion. Although this irregular chaotic motion is not directly visible on the macroscopic scale, it nevertheless manifests itself to us as "heat". It may also manifest itself as a mechanical pressure.

Thus, a kinetic theory of gases was developed, based on the assumption that a gas consisted of molecules in irregular or chaotic motion. As a first approximation, these molecules were thought of as having a small but finite size. But in a gas, this size was supposed to be much less than the mean distance between the molecules. Thus, the molecules move freely through space, except for occasional collisions. These collisions produce abrupt changes, both in the direction and in the magnitude of the velocity. It can be seen intuitively that in time a very irregular more or less random distribution of particle positions and directions of motion is to be expected, because of these collisions.

Let us now consider the effect of this irregular motion on the walls of the container. The walls will be struck almost continually, and each molecule will transmit a small momentum to the wall. The net effect will be similar to that of a rain of grains of sand. Because of the random distribution of the particles, an almost continuous pressure will be produced on the walls. It is this pressure which tends to cause a tank of compressed air to explode, and which provides the force that moves the piston in a steam engine or gasoline engine. Thus, we obtain an explanation of pressure in terms of the so-called kinetic theory of gases.

One of the strongest of the earlier experimental indications of the reality of random molecular motion came from a study of the Brownian motion. The botanist Brown discovered in 1824 that submicroscopic spore particles suspended in water exhibit an irregular and perpetual motion, without any visible source of energy. A similar behaviour was later observed with smoke particles suspended in air. For a long time efforts at explaining this phenomenon met with failure; but finally, after many hypotheses had been tried, it was shown that the Brownian motion could be explained both qualitatively and quantitatively as an effect of chaotic molecular motion. To do this, we first note that, although each smoke particle is small, it still contains of the order of 10^8 atoms or more. Thus, when it is struck by a molecule of the gas in which it is suspended, it will receive an impulse which causes it to change its velocity slightly. Now the gas molecules are moving quite rapidly (with velocity of the order of 10^4 cm./sec.), but because the smoke particle is much heavier than an atom, the result of its being struck by an individual atom will be a comparatively small change of velocity. Since it is being struck continually and in a very irregular way by the gas molecules, we expect to obtain a corresponding slow but irregular fluctuation in the speed of the smoke particle. The larger the particle, the less will be the fluctuation. Thus, some fluctuation in velocity will persist even for particles of macroscopic size (such as a chair), but its magnitude will be completely negligible. To obtain an appreciable effect we need to go to sub-microscopic bodies.

When the mean speed of fluctuation for particles of a given size was calculated, it was found to agree with that observed, within experimental error. Thus, the Brownian motion provided an important piece of evidence backing up the hypothesis of irregular molecular motions. Later, more direct evidence was found; for with modern techniques and apparatus it became possible to measure the velocities of individual atoms, and thus to show that they are really moving irregularly with the distribution of velocities predicted by the theory.

10. ON THE RELATIONSHIP BETWEEN MICROSCOPIC AND MACROSCOPIC LEVEL, ACCORDING TO THE MOLECULAR THEORY

The kinetic theory of gases described in the previous section was important, not only in itself but also because it was the first example within physics of a qualitatively new aspect of the laws of nature; viz. that the large-scale over-all statistical regularities can appear at the macroscopic level which are largely independent of the precise details of the complicated and irregular motions taking place at the atomic level. Because this kind of regular over-all statistical law has by now become quite common, not only in physics but also in many other fields, we shall give here a fairly detailed analysis of how such laws arise, for the case of the molecular theory of heat and the kinetic theory of gases.

Let us begin by considering a box of gas containing something of the order of 10^{23} molecules, each of which is moving in a very irregular path as a result of continual collisions with the other molecules. Clearly, to solve for the precise motions of each molecule would be a hopeless task. For, first of all, the problem is far beyond the domain of possibility, if only because of its sheer mathematical difficulty. But even if we could solve the mathematical problems involved we would be blocked by the practical impossibility of measuring the initial position and velocity of each molecule, which information is needed in order to make precise predictions in terms of the science of mechanics. And even if this information were available, it might not suffice, because our knowledge of the basic laws of mechanics themselves may not be perfect. Indeed, if one thinks with more care of the character of molecular motions, one sees that they possess an enormous instability. For example, a slight change of the initial angle of motion of any molecule will change appreciably its direction of motion after the first collision. This change would in turn lead to a still bigger change in the next collision, etc., and the cumulative effects of these changes would eventually carry the molecule in question to a very different region of space. Thus, the slightest error in any aspect of the theory, either the mathematics, or the knowledge of initial conditions, or the expression of the basic causal laws themselves, would in time lead to enormous errors in the predictions concerning the details of the molecular motions.

We see, then, that if we seek to make a detailed prediction of the behaviour of an individual system containing something of the order of 10^{23} molecules, we will be stopped on all sides. On the other hand, the very same factors that make such a detailed prediction impossible are also those that make possible a general prediction of the *over-all* or macroscopic average properties of the system without the need for precise information about exactly what the individual molecules are doing. We note that macroscopic average quantities (such as the mean number of molecules in a given region of space or the mean pressure on a given surface) are extremely *insensitive* to the precise motions and arrangements in space of the individual molecules. This insensitivity originates, at least in part, in the fact that an enormous number of different motions and arrangements in space can lead to practically the same values for these quantities. For example, if we interchange two molecules in space, we get a different set of microscopic conditions, but macroscopically the effects are the same. And if in a given region of space, a given molecule changes its direction of motion, the effects of this change on large-scale averages can be well compensated by suitable opposite changes in the motions of neighbouring molecules. Thus, irregular motions of many molecules will produce fluctuations, the effects of which tend, in the long run and on the average, to cancel out. Indeed, the above considerations are verified in quantitative detail in studies carried out in the science of statistical mechanics, which show that almost all possible initial conditions for the molecular positions and velocities lead to irregular motions in which the large-scale averages fluctuate very close to practically determinate mean values. Because these mean values depend almost entirely only on the general over-all properties of the molecules, such as the mean density, the mean kinetic energy, etc., *which can be defined directly at the large-scale level*, it becomes possible to obtain regular and predictable relationships *involving the large-scale level alone*.

It is clear that one is justified in speaking of a *macroscopic level* possessing a set of *relatively autonomous qualities* and satisfying a set of *relatively autonomous relations* which effectively constitute a set of *macroscopic causal laws*.* For example, if we consider a mass of water, we know by direct large-scale experience that it acts in its own characteristic way as a *liquid*. By this we mean that it shows all the macroscopic qualities that we associate with liquidity. For example, it flows, it "wets" things, it tends to maintain a certain volume, etc. In its motions it satisfies a set of basic hydrodynamic

* E.g. the laws of thermodynamics and macroscopic physics in general.

equations* which are expressed in terms of the large-scale properties alone, such as pressure, temperature, local density, local stream velocity, etc. Thus, if one wishes to understand the properties of the mass of water, one does not treat it as an aggregate of molecules, but rather as an entity existing at the macroscopic level, following laws appropriate to that level.

This does not mean, of course, that the molecular constitution of a liquid has no connection with its macroscopic properties. On the contrary, if we study the relationship between the two, we can even see *why* a relatively autonomous level is possible. The reason is just the insensitivity of the over-all large-scale behaviour to precisely what the individual molecules are doing. Part of this insensitivity comes from the cancellation of the effects of these individual motions described above. Another part comes from the effects of the inter-molecular forces. The forces between the molecules are such that they are in approximate balance when the liquid has a certain density. If the density is increased, so that the molecules are brought closer together, repulsive forces appear which automatically tend to bring the density back to its original value; while if the density is decreased, attractive forces between the molecules begin to manifest themselves; which likewise tend to bring the density back to its original value. Hence, there exists, in addition to the cancellation of the effects of the details of molecular motion, a certain *stability* of the characteristic modes of macroscopic behaviour, which tend to maintain themselves not only more or less independently of what the individual molecules are doing, but also of the various distur-bances to which the system may be subjected from outside.

The concept of relatively autonomous levels has been found to have a rather wide range of application. Thus, even in physics, it has been discovered that beneath the atomic level lies the level of the so-called "elementary particles" of physics, such as electrons, protons, neutrons, etc. And as we shall see later, especially in Chapter IV, there seems to be a new and as yet very poorly known level even below that of these elementary particles. In the other direction we have the molecular level (whose laws are studied mainly in chemistry, but partly in physics), the level of living matter (studied mainly in biology) which itself has many levels, as well as still other levels which the reader will easily think of.† In all these levels, however, we find the typical relative autonomy of behaviour, and the existence of sets of qualities, laws, and relationships which are characteristic of the level in question.

* The Napier-Stokes equations.
† As we shall see in Chapter V, this stratified structure may well be infinite, but it is not necessarily so.

11. QUALITATIVE AND QUANTITATIVE CHANGES

A very important further contribution of the atomic theory to the enrichment of the conceptual structure of physics was to permit a clarification of the relationship between qualitative and quantitative changes in terms of some simple examples which could be studied in considerable detail. As an illustration of how this was done, we shall discuss the transformation from gas to liquid to solid.

In the early phases of the development of physics, the qualitative properties associated with the gaseous, liquid, and solid phases of matter were simply assumed without further analysis. With the development of the atomic theory, however, it became possible to *explain* them, at least approximately, in terms of the *quantitatively* specifiable motions of the atoms and molecules of which the matter in question was constituted. The general lines of such an explanation are more or less as follows.

In the gaseous phase the molecules are, as we have seen, in perpetual and chaotic motion. Of course there exist attractive forces between the molecules, but the mean kinetic energy of random motion is so high that the molecules do not form stable combinations and, instead, more or less uniformly fill the entire space available to them. This is a characteristic property of the gaseous phase. As the temperature is lowered, the mean kinetic energy of the molecules decreases, and the inter-molecular forces begin to play a more important rôle. Thus, as we come near to the point of condensation of the gas, clusters consisting of a few molecules continually tend to be formed as a result of the mutual attractions of the molecules, but the mean kinetic energy is so high that these clusters break up almost as soon as they are formed. As the temperature is lowered further, the clusters tend to get larger; then, at a certain critical temperature, a qualitatively new phenomenon appears. Molecules continue to condense on a given cluster more rapidly than they leave. Thus, the clusters grow and become small droplets, after which the droplets later unite to form the new liquid phase. In this phase the substance does not fill the entire space available to it, but occupies a certain characteristic and definite volume, which is determined by the balancing of the attractive and the repulsive tendencies in the mutual interactions of all the molecules. Among the additional new qualities that appear in this phase is a relative incompressibility, the ability to "wet" surfaces, to dissolve various solids, and many others, all of which can be explained approximately by a more detailed analysis of the molecular motions, which we shall not, however, give here.

As the temperature of the liquid is lowered, further quantitative changes occur in its various properties (e.g. increase in density and in viscosity). These come from the decrease in mean molecular kinetic energy. Then, as the point of solidification is approached, incipient crystals, consisting of a small number of atoms arranged in regular and periodic lattices, begin to form. The tendency to form such crystals is again due to the inter-molecular forces, which are such that a lattice of this kind represents the most stable possible configuration. These crystals, however, break up almost as soon as they are formed, because of the disruptive effects of the random thermal motions. Below a certain critical temperature, however, the crystals begin to grow at a rate, on the average, faster than they break up, and the liquid is transformed into the qualitatively new phase of a crystalline solid. In this phase the substance tends not only to occupy a definite volume, but also to maintain a fixed shape, resisting efforts to deform it, and returning after it has been deformed to its original shape. Qualitative changes are also generally observed in many other properties (e.g. ability to transmit light, to polarize it, pattern shown on X-ray diffraction, etc.). As the temperature is lowered still further, additional quantitative changes in the properties of the crystal take place, coming from the continual decrease of the mean kinetic energy of vibration of the molecules around their mean positions in the crystal lattice.

We see, then, that *quantitative* changes in the mean kinetic energy of molecular motion lead to a series of *qualitative* changes in the properties of matter in bulk. These qualitative changes are generally foreshadowed as one approaches a critical temperature. As one passes such a critical temperature, however, two things happen. First, conditions are created in which completely different qualities come into being (e.g. the tendency in the case of the liquid phase to occupy a definite volume). Secondly, even those properties (such as specific heat, density, etc.) which are common to both phases show discontinuities in their quantitative behaviour as one passes through a transition point.

Let us now try to express more precisely just what is essential in a qualitative transformation. The most essential and characteristic feature of a qualitative transformation is that new kinds of causal factors begin to be significant in a given context, or to "take control" of a certain domain of phenomena, with the result that there appear new laws and even new kinds of laws, which apply in the domain in question. Thus, as we have seen, the volume of a gas is determined by the container, while that in the liquid phase is determined primarily by the inner conditions of the liquid itself. Hence, there appears a new quality; viz. a tendency to maintain a certain volume, which is reflected in a new form for the law relating the volume of the liquid to its other properties (such as temperature and pressure). Similarly, while the shape of a liquid is determined by that of the container, the shape of a solid is, like its volume, determined primarily by its own inner conditions. Thus, the new quality of rigidity appears, along with new laws that govern its characteristics.*

12. CHANCE, STATISTICAL LAW, AND PROBABILITY IN PHYSICS

Still another extremely significant new development of the conceptual structure of classical physics came from the introduction of the concepts of chance, statistical law, and probability in connection with the explanation of Brownian motion, the laws of thermodynamics, and other macroscopic properties. We shall therefore give here a brief account of how these concepts were applied in the field of atomic physics, for the purpose of calculating the statistical properties of large aggregates of atoms or molecules.

The modes of motion of atoms or molecules in such aggregates present all of the characteristic properties needed to give rise to chance fluctuations. Thus, as we have seen in connection with the discussion of levels given in Section 10, the details of the motions of individual molecules are (because of their extreme instability) very sensitive to exactly how the various other molecules are moving, and are also (because of collisions) extremely complicated and subject to very irregular and rapid fluctuations in their velocities, fluctuations which are effectively random.† As a result, the motions of each molecule are evidently contingent on an enormous number of rapidly fluctuating factors. Hence, *within the limited context in which we consider only a particular molecule*, it may be expected that a description of the motion as being subject to chance fluctuations should be substantially correct. Moreover, when we have to deal with a large number of molecules under uniform conditions (e.g. a box of gas), it is already evident, in qualitative terms at least, that in the long run and on the average each molecule will spend about the same fraction of time in any one region of space as in any other region of equal volume.‡ Thus, there is a uniform probability through-

* It should also be added that, as pointed out in Chapter I, Section 7, qualitative transformations satisfy many-to-one causal relationships that are independent of a wide range of variations in the detailed quantitative conditions accompanying the transformation.

† For a further discussion of the randomness of the velocities, see D. Bohm and W. Schutzer, *Nuovo Cimento*. (To be published.)

‡ A rigorous mathematical proof of this can be carried out for certain simple systems, but the mathematical difficulties have thus far been too great to permit a general treatment of this problem, which would be

out the whole box of occupying any particular region having a specified volume, regardless of where that region is. Finally, because of the randomness arising from the collisions, there will be no tendency in the long run for any one molecule to remain near any other specified molecule, so that each molecule has on the average a high degree of independence in its motions relative to that of the others.

Under these conditions, the theory of probability permits the calculation of a large number of properties of a statistical aggregate of molecules. The simplest example of the application of the theory is to note that an equi-probability distribution for the molecules implies that the mean or average density will be constant everywhere, a well-known property of a gas confined in a box under uniform conditions. Not only average properties, but also the mean *fluctuations away from the average*, can thus be predicted. This can be done with the aid of a well-known theorem, from which one concludes that, in this problem, the mean fractional fluctuation away from an average will be $\sqrt{n}$, where n is equal to the number of elements considered in obtaining this average. Thus, let us take a case of a gas with a mean density of 10^{17} molecules per cubic centimetre. If we consider a cube of side 10^{-5} cm. (with a volume of 10^{-15} cm.³), it will on the average contain 100 molecules. Because of the random motions of the molecules, however, this number will actually fluctuate rapidly with time; and according to the theorem cited above, the mean fractional fluctuation will be about 10. If we consider a cube of 10^{-3} cm. on a side, however, the mean number of particles in the cube will be 10^8, and the mean fractional fluctuation in this number will be only 1.0×10^{-4}. And with cube 10 cm. on a side, the mean number of molecules in the cube will be 10^{20}, while the mean fractional fluctuation will be only 1.0×10^{-10}. This example shows quantitatively how, as we consider the average containing more and more elements in a statistical aggregate, the chance fluctuations tend to cancel out more and more completely, thus creating the conditions for the appearance of a statistical law that grows more and more nearly determinate as the number of elements increases without limit.

A systematic application of the theory of probability to atomic physics is carried out in the study of statistical mechanics, which

equivalent to proving the so-called "quasi-ergodic theorem". (The conjecture that this theorem applies in typical problems of the kind described here is, however, made extremely plausible by the qualitative arguments that we have given here, as well as by the fact that such a theorem has in fact been proved for simple systems. See, for example, D. Bohm and W. Schutzer, *op. cit.*)

makes possible fairly precise calculation of a great many macroscopic properties of the system (e.g. entropy, heat capacity, equation of state, etc.) on the basis of the microscopic laws, and which also provides a model permitting a quantitative treatment of the way in which the macroscopic laws of thermodynamics arise out of the microscopic motions. It has also been applied in the study of Brownian motion, and in the study of the fluctuations of the macroscopic properties of matter near the critical points of liquids. Thus, the theory of probability has made an important contribution to our understanding of the relationship between microscopic and macroscopic levels by permitting us to take into account chance phenomena originating in the microscopic level without the need for either a precise and a detailed calculation of the motions of all the individual molecules in a large aggregate or a precise knowledge of the laws of the microscopic level.

13. THE ENRICHMENTS OF THE CONCEPTUAL STRUCTURE OF CLASSICAL PHYSICS AND THE PHILOSOPHY OF MECHANISM

We have seen in the preceding sections that, even apart from the development of the notion of fields, there occurred during the eighteenth and nineteenth centuries a number of very important additional steps that considerably enriched the conceptual structure of physics. These included the introduction of the concept of levels, that of quantitative changes that lead to qualitative changes, and that of chance fluctuations that tend towards approximately determinate laws for the mean behaviour of large aggregates. While, as we have already pointed out in Section 4, none of these concepts is in direct contradiction to a mechanistic philosophy, every one of them constitutes, in its general trend and spirit at least, a step away from the idea that there is an absolute and final fundamental law, which is purely quantitative in form, and which would by itself permit, in principle at least, the complete and perfect calculation of every feature of everything in the whole universe.

To see why these new concepts tend to lead away from mechanism we first recall that in the original form of the mechanistic philosophy, both the notion of qualitative changes and that of chance were regarded as nothing more than subjective aids to our thinking about the properties of matter *en masse*, so that they did not represent anything that was actually supposed to exist objectively in material systems. We have already seen in Section 11, however, that at least within the macroscopic domain, in a qualitative transformation, new qualities satisfying new laws become the significant and dominant causes in the domain in question. Moreover, the objective reality of the breaks in quantitative macroscopic properties, as well as that of the insensitivity of the qualitative change to quantitative details also cannot be denied. Similarly, it is evident (on the basis of the discussions given in Chapter I, Sections 8 and 9) that chance fluctuations exist objectively within specified contexts, and that the theory of probability provides a relatively precise mathematical expression of objective properties of these fluctuations, including the statistical regularities which arise on the basis of the cancellation of large numbers of chance fluctuations.

It is a most important characteristic of the mechanistic philosophy, however, that it permits one to make a limitless number of adjustments in his detailed point of view, without giving up what is essential to the mechanistic position. Thus, with the notion of qualitative changes, a great many physicists have effectively accepted the idea that these may well be objective (or at least as objective as anything else is). Nevertheless, they assert that such changes are not of fundamental significance, because they must, in principle at least, follow completely and perfectly in all of their details, in every respect, and without any approximation, from the quantitative laws of motion of the fundamental elements that make up the system, whatever these elements may be. Hence, it is maintained that qualitative changes are like passing shadows that have absolutely no independent existence of their own, but which depend for all their attributes on the quantitative laws governing the basic elements entering into the theory. It is evident that such an attitude implies also that the notion of a series of levels of law is likewise nothing more than a set of approximations to the absolute and final fundamental law, approximations in which the laws of the various levels depend completely for all their characteristics on the fundamental law, while the fundamental law has no dependence whatever on the laws of the various levels. Similarly, it is consistent with this point of view to suppose that chance and statistical laws arise out of nothing more than the sheer complexity and multiplicity of the motions of the basic entities entering into the fundamental causal law and that into the formulation of this latter law no element of chance whatever will appear.

We recall that historically the mechanistic philosophy was expressed in terms of the assumption that the basic units out of which the universe was supposed to be built are indivisible atoms. The purely quantitative laws governing the motions of these atoms were then regarded as the laws from which everything else followed.

It was discovered later, however, that the atoms are not really the fundamental units, because they are composed of electrons, protons, and neutrons in motion. From this fact one could already see that the assumption of the complete reducibility of everything in the whole universe to nothing more than the laws governing the motions of the atoms could not be perfectly valid, because the existence of an inner structure for the atoms permits the laws that govern them to be influenced by conditions existing at the large-scale level. Hence, the laws of the macroscopic level and those of the atomic level will actually be subject to *mutual* and *reciprocal* relationships.

As an example, let us consider the temperature of a substance. According to the atomic theory, this temperature is determined completely and perfectly by nothing more than the mean kinetic energy of the chaotic part of the molecular and atomic motions.* At temperatures of a few thousand degrees absolute, however, molecules and atoms dissociate into electrons and ions having qualitatively new properties compared with those of the undissociated systems, while at temperatures of the order of millions of degrees (existing in the interiors of stars), even the nuclei begin to transform, so that any atoms of a particular kind are always turning into atoms of other kinds. When these processes become important, the idea that the temperature is nothing more than a shadow-like property, determined solely by the mean kinetic energy of chaotic atomic and molecular motions, ceases to be able to represent the relevant features of the problem adequately. For an essential effect of the raising of the temperature is that the concept of the motions as reducible to displacements of atoms and molecules through space eventually becomes completely inapplicable. Hence, as long as we stay within the framework of the atomic theory, we are compelled to admit that large-scale properties such as the temperature may have a certain measure of relatively autonomous and independent being, in the sense that they contribute to the definition of the inner characteristics of the atoms, to the laws governing their external behaviour, and even to the conditions determining whether or not atoms of a given kind (or atoms of any kind at all) can exist. Under limited sets of conditions and for limited contexts, the effect of the large-scale laws on those of the atomic level are so small that we can ignore them, and then the notion of a complete determination of the macroscopic laws by those governing the atomic motions becomes a good approximation. But the original mechanistic assumption that this determination is not approximate and conditional, but instead perfect and absolute, is now seen not to be in accordance with the facts that have been discovered in the further progress of physics.

In answer to the above criticisms, the more modern mechanistic position is, of course, that the difficulties are due to the fact that the atoms are obviously not the fundamental elements, and that, instead, one must go to the electrons, protons, and neutrons, which are really

* See, for example, Section 9.

the fundamental ones. From the laws applying to these entities, we will then be able to deduce all the properties of the atoms, and to continue the deduction on to show that the laws of the macroscopic level also follow completely and perfectly from the same basis. Thus, we were right in our general goal, but wrong in prematurely supposing that we had reached it in the laws of the atomic level.

New developments in modern physics show, however, that this point of view cannot be correct, either. For the further progress of physics has disclosed that even the electrons, neutrons, and protons are not immutable, and that, under suitable conditions, they can be transformed into each other and into a whole host of qualitatively different kinds of particles called mesons, hyperons, etc. Such transformations take place when particles of very high energy collide with each other.* One can readily conceive of a macroscopic environment in which the temperature was so high that the mean particle energies were in the range in which these transformations could take place; and indeed, it is quite possible that such temperatures may eventually be produced artificially, or that they may even have existed naturally in the earlier phases of the development of the universe.† In such an environment, conditions at the large-scale level would significantly influence even the kinds of basic particles into which any system has to be analysed. As a result, the goal of deducing the laws of the higher levels completely and without approximation from those applying to electrons, protons, and neutrons, etc., proves to be unattainable, just as happened when the attempt was made to do this with the atoms as the fundamental units.

Of course, there is once again an easy way out of these difficulties, without leaving the framework of the mechanistic point of view. One merely needs to suppose that the really fundamental laws are not even the current ones applying to the motions of the electrons, protons, neutrons, mesons, etc., and that there must be a still more fundamental set of laws, which will finally settle the question, once and for all. But now the essentially philosophical assumption behind the mechanistic point of view has exposed itself clearly. For now one sees that not only are there no known cases of laws that accomplish the mechanistic aim, but even more, that even if we did have a law which seemed to explain everything that was known at a given time, we could never be sure that the next more accurate experiment or the next new kind of experiment would not show up some inadequacies that would lead eventually to a still more general and deeper

---

* These energies must be of the order of hundreds of millions of electron volts.

† This point is discussed in Chapter V, Sections 8 and 11.

set of laws. Indeed, this latter has been what has happened in physics thus far, with all the laws that have at one time or another been thought to be the final ones. Thus, the possibility will always be open that there will be a reciprocal influence between higher-level laws and those of any given lower level. This reciprocal influence may be negligible under familiar conditions, but very important under new conditions. The assumption that any given law is so fundamental that there is absolutely no reciprocal influence of this kind whatever is therefore one that cannot be proved on the basis of any conceivable kinds of experimental facts.

Similar conclusions can easily be obtained concerning the relationship of qualitative changes to the quantitative laws from which they can be predicted approximately. For we can readily see that there is a reciprocal influence of the qualitative state of matter on the quantitative laws which apply in any particular domain or level. Thus, for example, the precise form of the forces between molecules, which enter into the formulation of their laws of motion, depend in a fundamental way on the qualitative state of matter (e.g. is it a gas, a liquid or a solid,* etc.). Of course, we can understand this dependence approximately by considering the motions of the electrons, protons, and neutrons that constitute the atom; but, once again, the same basic problem arises. For it is possible to change the qualitative state of matter so much that even the basic quantitative properties of these latter particles will alter significantly. Thus, there are certain very dense stars, in which there are no such things, properly speaking, as atoms, but in which there are just dense masses of electrons, neutrons, and protons.† Under these conditions, there are good reasons to suppose that many basic quantitative properties of the electrons, neutrons, and protons should be quite different from what they are under more usual conditions.‡ Thus, we see that even the quantitative laws governing the electrons, protons, and neutrons depend somewhat on the qualitative state in which matter finds itself. Hence, we have not yet reached the goal of finding a purely quantitative law which completely and without approximation explains all

---

* For example, there may exist "many body" forces, which cannot be expressed as a sum of two body interactions, "exchange forces" resulting from the storing of electrons in a metal, directional forces resulting from the distortion of the atoms and molecules taking place when they are in a lattice, etc.

† The stars are so dense that the nuclei of the various atoms are practically in contact all the time.

‡ For example, because of the rapid exchange of mesons between these particles, the nuclear forces, magnetic moments, quadrupole moments, etc., are very probably quite different from what they are when the particles are present only in a low density.

qualities, so that the latter can have no possibility of making an independent contribution of their own to the expression of the laws of the whole system. Moreover, no experiment could possibly prove that a given set of quantitative laws *never* depends on the qualitative state of matter, since evidently, under new conditions not yet investigated, or in studies carried out to a higher level of approximation, such a dependence might eventually appear. Thus, the assumption that all qualitative changes are, at bottom, just passive "shadows" of quantitative changes of some basic set of entities, like the one that higher level laws are reducible completely to those of some fundamental level, cannot be founded on any conceivable kinds of experimental facts.

We can easily see that a similar result follows with regard to the attitude towards causality and chance that was characteristic of the mechanistic philosophy in the form that it had developed towards the end of the nineteenth century. Thus, as we have already pointed out in Chapter I, all causal laws known up to the present have been found to lead eventually to contingencies that are outside the scope of what can be treated by the causal laws in question. For example, as we saw in Section 2 of the present chapter, every mechanical law applies only to an isolated system, because its behaviour depends on boundary conditions that are determined in essentially independent systems external to the one under consideration. Even if we consider the entire universe as a single mechanical system, so that there is no outside,* then the same kind of problem arises. Thus, when we try to trace the causes of what happens at the macroscopic level with greater and greater precision, we eventually find dependence on the chance fluctuations of the essentially independent atomic motions. But these, in turn, depend in part on essentially independent chance fluctuations at the electronic and nuclear level (as well as on quantum-mechanical fluctuations which we shall discuss in Chapter III). These latter motions in turn depend in part on random fluctuations at still deeper levels, connected with the structure of the electrons, protons, neutrons, etc. (e.g. mesonic motions and probably even in a level below that of the elementary particles). Hence, there is no known case of a causal law that is completely free from dependence on contingencies that are introduced from outside the context treated by the law in question. Moreover, even if we had an example of a law that seemed to be completely free from such contingencies, the

---

* No meaning can be given to such a treatment in terms of predictions of actual experiments. It is, however, a useful philosophical abstraction to think of such a treatment, provided that we recognize that it merely serves as the basis for a discussion of certain important philosophical questions.

same general problem would arise as in the problem of reciprocal relationships between levels and between qualitative and quantitative laws. For the next step in scientific research might always disclose new factors existing outside the original context, on which the predictions of the laws in question were contingent. Thus, the notion that there is a final causal law completely free of contingency and from which all chance fluctuations can in principle be deduced completely and perfectly could not be based on any experimental facts.

In conclusion, we see that the mechanistic assumption that all the various levels, all qualitative changes, and all chance fluctuations will eventually be reducible completely, perfectly, and unconditionally to effects of some fixed and limited scheme of purely quantitative law, does not and cannot follow from any specific scientific developments. This assumption is therefore essentially philosophical in character. Whether it is desirable to make such an assumption will then be discussed in more detail in later chapters.

14. A NEW POINT OF VIEW TOWARDS PROBABILITY AND STATISTICAL LAW—INDETERMINISTIC MECHANISM

In response to the many difficult problems presented by the interpretation of chance phenomena and the associated statistical laws, there developed around the beginning of the twentieth century a new philosophical point of view towards these questions, which recognized the objective and fundamental character of chance and of the property of probability. This point of view eventually led, however, to the denial that determinate laws have any real significance, other than as approximations to the laws of probability which are valid when we are dealing with a statistical aggregate of things or processes.

The essential change brought in by this new point of view was the introduction of an element of arbitrariness into the theory. One still thought of the universe as a gigantic mechanical system with the property that everything in it can in principle be reduced completely and perfectly to nothing more than the results of purely quantitative changes taking place in suitable mechanical parameters. But instead of having its behaviour determined completely in terms of definite laws governing these parameters, this universal system could continually be subject to irregular alterations in the course of its motion. Since the parameters of the system are already assumed to describe everything that exists in the world, there is then evidently no place from which these irregular alterations in the motion could come. Thus, they could not have the character of ordinary chance fluctuations, which represent the effects of contingencies that cannot be taken into account in the context under discussion. Rather, they would represent a kind of fundamental and irreducible arbitrariness or lawlessness in the detailed behaviour of the world. Such a behaviour we may call by the name of "absolute chance", because it is not conceived of as being arbitrary and lawless relative to a certain limited and definite context, but rather as something that is so in all possible contexts.

The absolute arbitrariness and lawlessness in the detailed behaviour of individual phenomena is not assumed, however, to extend to a statistical aggregate. Instead it is supposed that the laws of nature can be expressed in terms of *probabilities*, which define, at least approximately, the long run and average behaviour that will be obtained in such statistical aggregations. All possible laws of nature are thus assumed to be expressible in terms of a set of purely quantitative relationships among the appropriate probabilities. For example, the notions of qualitative change and of relatively autonomous levels are still regarded as nothing more than approximate means of treating certain large-scale consequences of a basic and final fundamental law that is purely quantitative. But now this law is supposed to be probabilistic and not deterministic.

The point of view described above evidently renounces an important aspect of the various forms of the mechanistic philosophy that appeared from the sixteenth through the nineteenth centuries; namely, their determinism. But in doing this, it has conserved and in fact enhanced the central and most essential characteristic of this philosophy; namely, the assumption that everything in the whole universe can be reduced completely and perfectly to nothing more than the effects of a set of mechanical parameters undergoing purely quantitative changes. The fact that the details of these changes are completely arbitrary and lawless does not, however, make such a point of view essentially less mechanistic than the one in which these details are assumed to be determined by suitable properties of the system itself. Indeed, the introduction of absolute arbitrariness and lawlessness into a theory is analogous to taking as a model of the world, not an idealized frictionless machine of the type envisaged by Laplace but, rather, an idealized roulette wheel that would give an irregular distribution of results depending on nothing else at all (instead of on a multitude of factors lying outside the context available to people who play the game, as happens with real roulette wheels). The question of what constitutes a mechanistic philosophy, therefore, cuts across the problems of determinism and indeterminism. For this reason, we shall call the philosophy described in this section by the name of "indeterministic mechanism", to distinguish it from the deterministic mechanism which we have described previously.

Causality and Chance in Classical Physics: Philosophy of Mechanism

The nucleus of the indeterministic mechanist point of view towards chance is already presented in the work of von Mises* on the theory of probability. In this work, von Mises introduces the notion that in a genuinely random distribution of objects or events of the types to which we apply the theory of probability, there are no causal relationships at all, and that the distribution is completely "lawless". This means, however, that, whereas he admits that determinate laws can arise as approximations to the effects of laws of probability, which hold where large enough numbers of objects or events are involved,† he supposes that no analogous possibility exists by which laws of probability can arise as approximations to the effects of determinate laws. Thus, in this point of view, *laws of probability are regarded as having a more fundamental character than is possessed by determinate laws.*

There has been an extensive development of the point of view described above, but it has been carried to its logical conclusion only in connection with the usual interpretation of the quantum theory. We shall discuss this problem in more detail in the next chapter. Here, however, we may mention that the laws of the quantum domain are found to have a basically statistical character, which is such that, in general, they are expressed in terms of certain probabilities. At the large-scale level, these possibilities lead to practically determinate predictions; and in this way the familiar causal laws of classical mechanics emerge as statistical approximations. It is then assumed, however, as we shall see in more detail in the next chapter, that the probabilistic form of the current quantum theory can *never* be shown to be the result of an approximation to some deeper set of more nearly determinate laws. Thus, one is led to the conclusion that even the most fundamental laws of physics are, at bottom, nothing more than laws of probability, and that individual processes and events taking place in the atomic domain are completely lawless, in the sense suggested by von Mises.

We have thus come to an interesting inversion. For deterministic mechanists regard chance as reducible completely and perfectly to an approximate and purely passive reflection of determinate law. On the other hand, indeterministic mechanists such as von Mises and the proponents of the usual interpretation of the quantum theory, regard determinate law as reducible completely and perfectly to an approximate and purely passive reflection of the probabilistic relationships associated with the laws of chance.

* R. von Mises. Wahrscheinlichkeit, *Statistik und Wahrheit*, dritte Auf. Wien, Springer (1951).

† E.g. the laws determining the pressure produced by a large number of molecules.

Now, in connection with the indeterministic mechanist point of view, a basic question that must be settled is whether the details of chance fluctuations are ever really completely arbitrary and lawless relative to *all possible* contexts. In answer to this question, we first remark that in a very wide range of applications of the concept of chance it has actually been possible, as we pointed out in Chapter I, Sections 8 and 9, to show at least *qualitatively* that by broadening the context sufficiently we find more and more nearly unique causal relationships applying within the chance fluctuations. Moreover, in many cases, one can even demonstrate the same conclusion *quantitatively*. For example, it has been proved mathematically that there exists a wide class of determinate sequences involving complicated chains of events or events determined by a large number of independent causal factors, which possess, to an arbitrarily high degree of approximation, the essential statistical properties that are characteristic of distribution treated in terms of the theory of probability.* Thus, one sees that the possibility of treating causal laws as statistical approximations to laws of chance is balanced by a corresponding possibility of treating *laws of probability as statistical approximations to the effects of causal laws.* It follows from this, however, that the original notion of von Mises that the laws of probability apply to a "completely lawless" distribution of objects or events can never be given a clear meaning in any specific problem or application. For the possibility is always open that any given set of laws of probability applying within a given context will eventually be seen to be approximations to new kinds of causal laws applying in broader contexts.

The assumption that any particular kind of fluctuations are arbitrary and lawless relative to *all possible* contexts, like the similar assumption that there exists an absolute and final determinate law, is therefore evidently not capable of being based on any experimental or theoretical developments arising out of specific scientific problems, but it is instead a purely philosophical assumption. The question of whether it is desirable to make such an assumption we shall discuss in the next chapters.

15. SUMMARY ON MECHANISM

We have seen that the philosophy of mechanism which started out with such brilliant prospects during the time of Newton, ran into a series of difficult problems that began to become especially serious

---

* See, for example: H. Weyl, *Ann. der Mathematik*, **77**, 333 (1916); H. Steinhaus, *Studia Mathematica*, **13**, 1 (1953); G. Klein and T. Prigogine, *Physica*, **19**, 74, 89, and 1053 (1953); D. Bohm and W. Schutzer, *Supplemento al Nuovo Cimento*, Vol. II, Series X, n. 4, p. 1004 (1955).

during the nineteenth century. These problems were resolved by means of a series of successive accommodations and modifications, which retained, however, the essential characteristic of assuming that, in principle, everything would finally be reducible completely and perfectly to an ultimate set of purely quantitative laws, involving perhaps bodies alone, perhaps bodies and fields, or perhaps fields alone. The various qualitative changes occurring in matter as well as the existence of various levels would then one day be seen to be merely a result that follows completely and perfectly, in principle at least, from the fundamental quantitative laws.

We have seen, however, that this point of view does not very well fit the experimental facts that are available up to the present. For further progress in physics has shown that all the various purely quantitative theories that were at different times thought to be the fundamental ones are actually approximations to still deeper and more general theories containing qualitatively new types of basic entities that are related by correspondingly new types of laws. Moreover, the possibility will always be open that, as has so often happened already, future experimental results may show the need for still further changes of a far-reaching character in our basic theories. As a result, there is no conceivable way of proving that the laws of the various levels and of qualitative changes are *completely* and *perfectly* reducible to those of any given quantitative theory, however fundamental that theory may seem to be.

The problem of probability and chance proved to be a particularly difficult one for a mechanistic philosophy. For, besides leading to problems very similar to those that arise in connection with the concepts of levels and of qualitative changes, it brought the mechanists on to one horn or the other of a dilemma; namely, the need to decide, once and for all, and without any possibility of experimental proof, whether determinate law is the fundamental category, while chance and probability are only passive reflections, or whether chance and probability are fundamental and determinate law is only a passive reflection.

It should be noted, however, that all the new developments that occurred in the conceptual structure of physics during the nineteenth century were in such a sense as to suggest that none of the various possible mechanistic schemes, deterministic and indeterministic, that have been suggested at different times are really fundamental, but that rather, what should be our fundamental starting-point is the full richness of the patterns of natural law described in Chapter I. This pattern implies that all the laws of the various levels and all the different general categories of law, such as qualitative and quantitative, determinate and statistical, etc., represent different but necessarily interrelated sides of the same process. Each side gives an approximate and partial view of reality that helps correct errors coming from the sole use of the others, and each treats adequately an aspect of the process that is not so well treated or perhaps even missed altogether by the others. Within the framework of this general pattern, one can quite easily integrate all the new developments in physics that we have described here, and a great deal more besides. Thus, there is no need to make continual assumptions that certain types or categories of law are the final ones from which everything else follows completely and perfectly, assumptions which can never be proved experimentally and which are always subject to being disproved with the further progress of science. Nor do we have to be faced with unresolvable dilemmas, such as that of making a final decision without any possible experimental proof whether deterministic or probabilistic laws are really the fundamental ones. We recognize the contribution to our understanding of nature made by every concept and by every category of law, and we leave for further scientific research the problem of finding out the extent to which any one concept or category of law can, within some degree of approximation and under some conditions, be shown to follow necessarily from any other specified set of concepts and categories of law.

After pursuing further the course of development of the indeterministic form of the mechanist philosophy in the quantum theory in Chapter III, and proposing a new interpretation of the quantum theory in Chapter IV, we shall then return in Chapter V to this problem, to give a more detailed exposition of how modern physics fits into the above-mentioned general pattern of natural law.

CHAPTER THREE

The Quantum Theory

1. INTRODUCTION

IN the previous chapter we have given a discussion on the evolution of classical physics, starting with Newton's laws of motion, and continuing on to all the new developments which occurred up to the end of the nineteenth century. Throughout this time, however, the general philosophical view held by physicists was that of deterministic mechanism. For it was felt that even though the *details* of the theories that were then current would eventually have to undergo various modifications in response to the results of further experiments, the basic general scheme in which all theories are formulated in terms of differential equations determining the future behaviour of everything in the universe completely in terms of their states at a given instant of time would never have to be changed. For example, Lord Kelvin, one of the leading physicists of the time, expressed the opinion that the basic general outline of physical theories was pretty well settled, and that there remained only "two small clouds" on the horizon, namely, the negative results of the Michelson-Morley experiment and the failure of Rayleigh-Jeans law to predict the distribution of radiant energy in a black body. It must be admitted that Lord Kelvin knew how to choose his "clouds", since these were precisely the two problems that eventually led to the revolutionary changes in the conceptual structure of physics that occurred in the twentieth century in connection with the theory of relativity and the quantum theory.

Now, while the theory of relativity brought about important modifications in the specific forms in which the causal laws are expressed in physics, it did not go outside the previously existing theoretical scheme, in which the values of suitable parameters at a given instant of time would in principle determine the future behaviour of the universe for all time. We shall, therefore, not discuss the theory of relativity in this book, in which we are interested

68

primarily in the question of causality, because this theory raised no question that went to the root of the problem of causality.

On the other hand, the quantum theory had, from the point of view of a discussion of causality, an effect that was much more revolutionary than that of relativity. Indeed, it was the first example in physics of an *essentially statistical theory*. For the quantum mechanics did not start from a treatment of the laws of individual micro-objects and then apply statistical considerations to those laws, as is done in classical mechanics (see Chapter II, Section 14). Rather, from the very beginning, it took the form of a set of laws which gave in general only statistical predictions, without even raising the question as to what might be the laws of the individual systems that entered into the statistical aggregates treated in the theory. Moreover, as we shall see, the indeterminacy principle of Heisenberg led physicists to conclude that in investigations carried out to a quantum-mechanical level of accuracy no precise causal laws could ever be found for the detailed behaviour of such individual systems, and thus they were led to renounce causality itself in connection with the atomic domain.

We shall see, however, that the indeterminacy principle necessitates a renunciation of causality only if we assume that this principle has an absolute and final validity (i.e. without approximation and in every domain that will ever be investigated in physics). On the other hand, if we suppose that this principle applies only as a good approximation and only in some limited domain (which is more or less the one in which the current form of the quantum theory would be applicable), then room is left open for new kinds of causal laws to apply in new domains. For example, as we shall see, there is good reason to assume the existence of a sub quantum-mechanical level that is more fundamental than that at which the present quantum theory holds. Within this new level could be operating qualitatively new kinds of laws, leading to those of the current theory as approximations and limiting cases in much the same way that the laws of the atomic domain lead to those of the macroscopic domain. The indeterminacy principle would then apply only in the quantum level, and would have no relevance at all at lower levels. The treatment of the indeterminacy principle as absolute and final can then be criticized as constituting an arbitrary restriction on scientific theories, since it does not follow from the quantum theory as such, but rather from the assumption of the unlimited validity of certain of its features, an assumption that can in no way ever be subjected to experimental proof.

We see, then, that in certain respects twentieth-century physicists have continued the classical tradition of conceiving of the *general* features of their theoretical schemes as not subject to future modifications in new domains or in more accurate investigations of already known domains. But these general features do not fit into a deterministic mechanist scheme, but rather into an indeterministic mechanist scheme. However, the indeterministic mechanism takes a more subtle form than it had in the earlier versions described in Chapter II. Thus, although a great many of the proponents of the usual interpretation of the quantum theory have had the express purpose of going outside the limits of a mechanistic philosophy, what has actually happened has just been a switch from deterministic to indeterministic mechanism.

2. ORIGIN OF THE QUANTUM THEORY*

The first evidence in favour of the quantum theory came from the work of Planck and Einstein. Let us recall that classical physics was characterized by the assumption that the bodies of which matter was composed moved continuously and exchanged energy continuously with the electromagnetic waves such as those of light, which we discussed in the previous chapter. On the other hand, Planck and Einstein, studying certain experiments in which matter exchanged energy with light, came to the conclusion that the light transmits energy to matter in the form of "quanta" or bundles, of size $E = h\nu$ where ν is the frequency of the light wave and h is a universal constant, which was later called Planck's constant.

Let us now consider some of these experiments in more detail. To do this, it is necessary first to discuss classical theory a little further. According to the evidence of interference and diffraction, which we discussed in the previous chapter, light consists of waves. With the aid of Maxwell's equations, and experiments such as those of Hertz, strong evidence was obtained in favour of the conclusion that these waves are electromagnetic in nature. Now, just as a water wave can be created by a body disturbing the surface of the water, when it moves up and down, an electromagnetic wave can be produced when a charged particle, such as an electron, moves through space with oscillatory motion, and thus disturbs the electric and magnetic fields. In both cases the wave spreads out continuously. In the case of electrons, this motion can create light-waves, radio waves, or other types of electromagnetic waves, depending on the frequency. Water waves can set floating objects in oscillatory motion with an energy proportional to the intensity of the wave. Similarly, with light-waves, the electromagnetic fields will act on charged particles such as electrons and impart to them an oscillatory motion, with an energy proportional to the intensity of the light-wave.

* For a more detailed discussion of this subject, see D. Bohm, *Quantum Theory*, Prentice Hall, New York, 1951, Part I.

This theory was tested experimentally by studying the photoelectric effect. In such a study, light shines on a metal surface, A, placed in a glass tube that has been evacuated (see Fig. 2). Previous experiments had shown that a metal contains electrons in large numbers. Thus, it would be possible for electrons occasionally to be liberated from the illuminated metal surface. A plate, B, is therefore placed in the tube to collect any electrons that may be liberated from the illuminated surface, A. These electrons would give rise to an electric current, which can be measured by means of a galvanometer, G.

The first result of the experiment is that when the plate, A, is illuminated, a current is actually observed in the galvanometer, thus demonstrating the liberation of electrons by light from the metal surface. The next step is to measure the energy of these electrons. This is done by establishing an electric potential difference between A and B, in such a direction as to tend to turn the electrons around before they reached the plate, B. Thus, as the potential difference

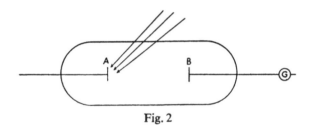

Fig. 2

was increased, more and more electrons are turned around; and at a certain critical value the current becomes zero. This critical value is clearly equal to the *maximum* kinetic energy with which electrons are liberated. A more careful analysis of the way in which the current varies with potential permits one to deduce the distribution of kinetic energies of the electrons. An analysis of the data from such an experiment established the following result:

When the plate, A, is illuminated with light of frequency, ν, the electrons all gain the *same* amount of energy, $E = h\nu$, which depends only on the frequency of the light, but not on its intensity. Thus, when the light is very weak, the electrons still gain the same energy, $E = h\nu$, but correspondingly fewer electrons are liberated.

This result is in clear contradiction to the predictions of classical theory which state that the energy gained should depend *continuously* on the intensity of the radiation, so that with weaker light less energy should be gained. Indeed, taken *by itself* this experiment would suggest that light does not consist of waves, but of swarms of

71

small discrete particles, of energy, $E = h\nu$, so that when one of the particles strikes an electron, it can transfer the discrete energy, $E = h\nu$. A weak light-wave would have few such particles and an intense wave would have many. This model had, in fact, already been suggested earlier by Planck on the basis of a study of the distribution of energy emitted by a heated black body. Planck had shown that classical theory led to a quite definite prediction for this energy distribution, which was wrong. But the assumption that energy comes in packets, or quanta with energy $E = h\nu$, explained these experiments very nicely. The same constant, h, that was needed in Planck's theory also predicted correctly the energy of electrons in the photo-electric effect. Thus, the evidence was very strong that light energy really comes in the form of quanta.

On the other hand, the evidence that light consists of discrete particles comes into conflict with the experiments on interference, which seem to demand that light is a *continuous* form of wave motion. Consider, for example, the experiment discussed in the previous chapter in which a beam of light is incident on two slits. If only the first slit is open, we get a certain more or less uniformly varying pattern of light on the screen. But if another slit, B, is opened, we get a set of alternate bands of light and darkness. Thus, the opening of the second slit, B, can create darkness at certain points, where with one slit alone there was light. This was explained by assuming that the motions due to the waves coming through the second slit could at certain points cancel those due to the waves coming through the first, thus producing darkness. But if light consists of a swarm of particles, then the opening of the second slit should in general be expected to *increase* the amount of light reaching each point on the screen, and should, at least, never *decrease* it.

At first sight, it might be thought that this phenomenon could perhaps be explained by supposing that light consists of a swarm of quanta which interact with each other so that when two slits are open, the paths of the particles of light would be modified in such a way that they could not arrive at the dark parts of the fringes. But later experiments done by Vavilov, using light so weak that only one quantum could enter the apparatus at a time, showed that this explanation is not tenable. For in this case each individual quantum liberates a single electron. But after a large number of quanta have passed through the system, each separately and independently, there will appear a statistical pattern in the locations of the points from which electrons were liberated; and this pattern will approach the classical pattern of fringes of light and darkness. Thus, the opening of a second slit can prevent a *separate* and *independent* quantum from reaching certain dark points in the pattern which it could reach if that slit were closed. Hence *even an individual quantum shows some wave-like properties.* On the other hand, it also shows some particle-like properties, not only because it gives up to an electron a field quantum of energy, $E = h\nu$, but also because in a beam containing a small number of quanta *there are statistical fluctuations in the time and place of liberation of an electron which are just those that would come from a beam of particles distributed in space in a highly irregular or "random" way* (such as one would expect to have if the particles were emitted by some source undergoing chaotic molecular motion).

We seem to be faced with a paradox. One set of experiments suggests that light is a form of *wave motion,* while another suggests equally strongly that it consists of discrete *particles* or *quanta.* How this paradox is to be resolved will, however, be discussed later. For the present, we shall continue our presentation of the development of the quantum theory.

The next step was due to Bohr. There had already developed an extensive body of investigations, leading to the conclusion that matter is made of atoms, and that these atoms are in turn made of light negatively charged particles called electrons which circulate around a heavy positively charged nucleus, in much the same way as the planets circulate around the sun. As the electron goes around the nucleus, it should emit electromagnetic waves of the same frequency as that of rotation. This frequency could be calculated, and was found to be of the order of 10^{15} cycles per second, which is of the order of that of light. Thus, one could explain qualitatively how light is emitted by matter.

When the process of emission of light was studied in more detail, however, a number of serious contradictions between the existing theory and experiment were discovered. The most striking of these arose out of the mere fact, at first sight almost trivial, that atoms exist stably. For according to classical theory, a moving charged particle such as an electron should lose energy by radiating electromagnetic waves at a rate which was shown by calculations based on Maxwell's equations to be proportional to the square of the acceleration of the electron. An electron moving in a curved orbit is always being accelerated towards the centre of the atom. Thus, it should continually be losing energy. This energy can come from only one source, the potential energy of attraction of the nucleus for the electron. But for this potential energy to be liberated the electron must fall towards the nucleus. Thus, we predict that the electron will move in a spiral orbit, and will reach the nucleus in a time that calculation shows to be an extremely small fraction of a second. However, what happens in reality is that the electron stops radiating when it reaches a certain normal radial orbit characteristic of the atom in question,* in which it remains indefinitely thereafter as long as it is not disturbed. Hence, some new factor must be present, not contained in classical theory, which explains why the electron stops radiating when it reaches the normal radius of the atom.

Another important contradiction between classical theory and experiment arose in a detailed study of the frequencies of the radiation emitted by atoms. According to classical theory, there should exist a continuous range of possible sizes of orbits of the electrons. And since each different size of orbit led in general to a different frequency of revolution of the electron around the nucleus, there should be possible a corresponding continuous range of frequencies of the light emitted. Indeed, because of the chaotic character of motion at the atomic level, a given sample of matter, such as a tube of hydrogen gas, should contain atoms with a chaotically distributed range of sizes of orbits which, because there are so many atoms (10^{20} or more), would appear practically continuous. Thus, a con-

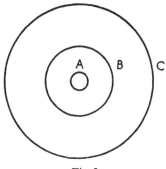

Fig. 3

tinuous range of frequencies of light should be emitted. In reality, however, only certain discrete frequencies are obtained experimentally.

Bohr analysed this problem very carefully, and finally was able to resolve the above contradictions with experiment (as well as a number of others which we have not mentioned here) by means of a totally new kind of hypothesis. He postulated that the continuous range of orbits permitted by classical theory were not in reality possible, and that the electron could follow only certain discrete (i.e. quantized) orbits, such as those illustrated in Fig. 3. By postulating that among these there existed a smallest possible orbit (indicated by A in the figure) with a lowest possible energy, he immediately

* Experiments had already indicated that this size is of the order of 10^{-8} cm.

explained the stability of atoms. For once the electron entered this orbit, it would not be able to lose any more energy, because no more orbits of lower energy would be available for it to go to. Thus, it would remain in this orbit until it was disturbed from outside.

If, for one reason or another, the electron were in an orbit, for example, C, with an energy E_C, higher than that in the bottom orbit, then Bohr postulated that it could jump from that orbit to a lower one, for example, B, radiating the *full* energy difference $E_C - E_B$, in one single quantum of light, with a frequency given by the Einstein relation $E_C - E_B = h\nu$. This postulate clearly has as a consequence that only discrete frequencies of light can be emitted, corresponding to the discrete jumps between the various possible energy levels.

Having resolved the contradictions between theory and experiment in a *qualitative* way, Bohr proceeded to derive a *quantitative* rule, permitting him to calculate the allowed energy levels and the corresponding frequencies of light emitted at first for hydrogen atoms, and later for a few other simple kinds of atoms. This quantitative rule permitted for these cases a prediction of the frequencies emitted with a very high order of precision. These predictions enormously increased the plausibility of the theory; for they involved such a large number of frequencies and reached such a high level of precision that it would have been difficult to believe that the agreement was a coincidence.

Thus Bohr had presented very convincing evidence in favour of the idea that not only does the energy of light come in discrete packets or quanta, but also that of electrons. Further investigations, which we shall not discuss here, established a similar discontinuity in *all* forms of energy. In other words, a basic "atomicity" of energy in general had been disclosed. The size of the basic units was, however, not the same under all possible conditions. For example, with light it was proportional to the frequency, but in atoms it depended on more complex rules.

It must be emphasized, however, that it had not been explained *why* the energy is atomic in character. The atomicity had just been *postulated*; and on the basis of this postulate, many properties of atoms and of radiation were explained, which had been in contradiction with the conclusion drawn from classical physics that the energy can vary in a continuous way. Moreover, no explanation was offered for the process by which a quantum was emitted and absorbed, during the course of which the electron obviously had to jump from one discrete orbit to another. At this level of the theory, it was merely accepted that these processes occur somehow, in a way that it was hoped would be understood better later (as it was also hoped in connection with the problem of the very existence of discrete orbits).

A first step towards a better understanding of the discrete energies of atomic orbits was made by de Broglie. De Broglie's starting-point was in the suggestion that just as light-waves had a particle-like character, atomic particles might also have a wave-like character. In doing this, he was guided by the appearance in connection with many different kinds of classical waves of sets of discrete *frequencies*. For example, a string fixed at each end must vibrate in integral multiples or "harmonics" of a certain "fundamental" frequency, determined by the length, density, and tension of the cord. Likewise, sound waves in a box can have only discrete frequencies, but these are in a more complicated relation than just that of integral multiples. In general, whenever a wave is confined to oscillate within a definite space, it may be shown to have discrete possible frequencies of oscillation.

De Broglie then postulated that there exists a new kind of wave connected with an electron. As to the precise nature of this wave, most of its properties will not be important at this level of the theory. What is important here is that if it is confined within an atom, it will have discrete frequencies of oscillation. If we now postulate that the Einstein relation, $E = h\nu$, connecting the energy of the wave to its frequency applies to these waves just as it applies to light-waves, then the *discrete frequencies* will imply *discrete energies*.

The next step was to put this qualitative theory into a more quantitative form. De Broglie did this by showing, on the basis of arguments coming from the theory of relativity, that the Einstein relation, $E = h\nu$, led to another relation, $p = h\lambda$, connecting the wave-length, λ, of these waves with the momentum, p, of the electron. When the wave-length, λ, was evaluated for an electron of a typical momentum encountered under usual experimental conditions, it was found to be of the order of atomic dimensions. Now, from experience with light and other types of waves, we already know that a wave-like character is manifested clearly only when the wave meets obstacles that are not too much larger than a wave-length in size; otherwise it goes in a practically straight line as if it were a particle. Thus at the large-scale level, de Broglie's waves would not show themselves clearly, and the electron would act as if it were nothing but a classical particle. At the atomic level, however, the wave connected with the electron would produce important new effects. Among these would be the appearance of discrete frequencies of vibration resulting from confinement of the waves within an atom. Using the relation which he had discovered ($p = h/\lambda$) and the Einstein relation, $E = h\nu$, de Broglie was then able to calculate both the frequencies and the corresponding energies of the discrete possible modes of vibration of these waves; and out of these calculations he obtained exactly the same energies as those coming from Bohr's theory. Thus, the Bohr energy levels were explicable in terms of an assumed wave, *provided that one also assumed that the energy of this wave was related to its frequency by the Einstein relation, $E = h\nu$.*

Later experiments done by Davisson and Germer on the scattering of electrons from metallic crystals disclosed a statistical pattern of strong and weak scattering very similar to the fringes obtained by passing a beam of light quanta through a set of slits. The idea was then suggested that perhaps here the waves postulated by de Broglie were manifesting themselves, and that the regular array of atoms in the crystal was playing the rôle which the set of slits plays in optical interference experiments. When the length of the assumed wave was calculated on the basis of the observed pattern of strong and weak scattering, it was found to agree with that obtained from de Broglie's theory. Thus, the conjecture that electrons have some wave-like properties received a brilliant experimental confirmation. Later, similar experiments showed that other particles, such as protons, molecules, neutrons, etc., have similar wave-like properties, which also satisfy the de Broglie relations. Thus, by now we have the point of view that *all matter* has such wave-like properties.

Meanwhile, the wave theory of de Broglie had been developed into a much more precise form by Schrödinger, who obtained a partial differential equation for these waves, which determines their future motions in much the same way that Maxwell's equations determine the future motions of waves in the electromagnetic field. Schrödinger's equations permitted the precise calculation of energy levels in a very wide variety of atomic systems, which it was not possible to treat either by Bohr's theory or by de Broglie's theory; and such calculations led to a very impressive agreement with experiment in all cases. Moreover, the Schrödinger equation permitted a continuous treatment of how the wave moves in a transition from one allowed energy level to another, and thus led to the hope that perhaps the mystery of how a transition between allowed energy levels takes place could now be solved.

At this point new and apparently somewhat paradoxical limitations on the wave theory were discovered. For Schrödinger originally proposed that the electron should be thought of as a continuous distribution of charge. The density of this charge he assumed was related to the wave amplitude ψ, by the relation, $\varrho = |\psi|^2$. Thus, the waves of de Broglie and Schrödinger were to be interpreted as waves of electric charge. In favour of this suggestion, if the electric charge were assumed to be related to the wave amplitude in this way, then Schrödinger's equation led automatically to the conclusion that the total amount of charge would remain constant, no matter how it flowed from place to place (i.e. it would be conserved), thus demonstrating a consistent feature of the interpretation.

Unfortunately, however, the interpretation was tenable only as long as the Schrödinger wave remained confined within an atom. In free space, a simple calculation showed that, according to Schrödinger's equation, the wave must spread out rapidly over all space without limit. On the other hand, the electron is always actually found within a comparatively small region of space, so that its charge density clearly cannot in general be equal to the value, $\varrho = |\psi|^2$, postulated by Schrödinger.

To deal with this problem, Born proposed that the wave intensity represents not an actual charge density of the electron, but rather the *probability density* that the electron, conceived of as a small localized particle, shall be found at a certain place. Thus, the fact that the wave amplitude for a free electron spreads out over all space is no longer in contradiction with the appearance of the electron itself at a certain place. The conservation of $|\psi|^2$ can then be interpreted in terms of the fact that the total probability that the particle can be found somewhere in space must remain equal to unity with the passing of time.

It was not experimentally feasible to verify Born's hypothesis *directly* by observing the locations of particles in a statistical aggregate, but it was possible to verify it *indirectly*. Thus, in a transition between allowed energy levels, the change of the Schrödinger wave from one mode of vibration to another now had to be interpreted in terms of a continuously changing probability that the electron had one energy or the other.* Thus, it was possible to calculate *probabilities of transition* between energy levels under various conditions; and these probabilities were found to be in agreement with experiment. So much indirect evidence in favour of Born's hypothesis has by now been accumulated that physicists generally accept Born's interpretation of the Schrödinger wave function, ψ, as being correct.

Let us now sum up the results that had been obtained thus far:

(1) Energy in general appeared to have a certain atomicity, both in the form of light quanta and in the form of discrete allowed energy levels for matter.

(2) All manifestations of matter and energy seemed to have two

* Hence, the problem of describing what actually happens in an individual transition process had not yet been solved. With the interpretation of Born, the Schrödinger wave only treated the mean behaviour in a statistical ensemble of cases.

possible aspects, that of a wave and that of a particle. The numerical value, E, of the energy in the particle-like manifestations was always related to the frequency, ν, in the wave-like manifestations by the Einstein relation, $E = h\nu$. The numerical value, p, of the momentum in the particle-like manifestation was likewise related to the wavelength, λ, by the de Broglie relation, $p = h/\lambda$.

(3) The basic laws of atomic physics appeared to be statistical in form. Thus, the Schrödinger wave function permitted in general only the prediction of the probability that a certain result could be obtained in an observation of an atomic system. Likewise, experiments with beams of light quanta (as well as a subsequent detailed theory of quantum electrodynamics) showed that one obtained random statistical fluctuations in the times and places at which photo-electrons were liberated, and that, in general, only the probability of such a process seemed to be predictable.

Nevertheless, *some* properties of the individual systems could be predicted with certainty, for example, the energy levels.

3. ON THE PROBLEM OF FINDING A CAUSAL EXPLANATION OF THE QUANTUM THEORY

At this point, physicists faced a very difficult problem. A number of rather puzzling general properties of matter had been found, including a peculiar combination of wave and particle-like properties that seemed very difficult to explain, as well as a very strange combination of determinate and statistical aspects of a type that had never been met before. Nevertheless, although the new phenomena were very unusual, it was by no means true that they suggested no hints at all as to how they might be given a causal explanation.

In order to help clarify the position eventually adopted by most modern theoretical physicists with regard to the interpretation of the quantum theory, as well as to clarify the criticisms of this position that we shall make later in this chapter, we shall give here a very brief sketch indicating in a general way a possible line of research along which one could have sought a causal explanation of the quantum theory. More detailed proposals with this end in view will, however, be discussed only in the next chapter. The main purpose of the discussion here will be merely to help bring out more clearly the full implications of the usual interpretation of the quantum theory by introducing an opposing point of view, as a sort of counterfoil, which serves to show what the usual point of view denies.

To show the lines along which one could have sought such a causal explanation of the quantum theory, let us begin with point (3), the appearance of a peculiar combination of statistical and individual laws. Hitherto in physics (as in other fields) when one had met with an irregular statistical fluctuation in the behaviour of the individual members of an aggregate, one assumed that these irregular fluctuations also had causes, which were however as yet unknown to us, but which might in time be discovered. Thus, in the case of the Brownian motion, the postulate was made that the visible irregular motions of spore particles originated in a deeper but as yet invisible level of atomic motion. Hence, *all* the factors determining the irregular changes in the Brownian motion were not assumed to exist at the level of the Brownian motion itself, but rather, most of them were assumed to exist at the level of atomic motions. Therefore, if we study the level of Brownian motion itself, we can expect to treat, in general, only the statistical regularities, but for a study of the precise details of the motion, this level will not be complete.

Similarly, one might suppose that in its present state of development, the quantum theory is also not complete enough to treat all the precise details of the motions of individual electron, light quanta, etc. To treat such details, we should have to go to some as yet unknown deeper level, which has the same relationship to the atomic level as the atomic level has to that of Brownian motion. Of course, some of the properties at the atomic level are determinate, but this creates no difficulties of principle, since it is quite possible that the factors determining just these properties can be defined in the atomic level alone, while these determining other properties cannot. Thus, we can understand why the atomic theory has, in general, to deal with probabilities, even though it can predict some properties of individual systems.

Once we admit that the entire conceptual framework of the existing quantum theory may not be adequate for a treatment of all the detailed properties of individual systems, then an unlimited number of new possibilities open up, since the as yet unknown properties of the deeper level are completely at our disposal. For example, we can already see in a qualitative way how the Einstein relation, $E = h\nu$, connecting the frequency of an oscillation with its energy might be explained. Thus, even in classical physics, many examples of oscillations satisfying *non-linear equations** were known, in which there were only certain discrete stable frequencies of oscillation, and in which the energy was related to the frequency in a definite way (e.g. the motions of the electron in a synchro-cyclotron†). It is true that

* A linear equation is one having the property that the sum of two of its solutions is also a solution of the same equation. A non-linear equation does not have this property. For this reason it is much more difficult to treat mathematically, as no such simple relations exist, in general, among its solutions.

† See, for example: D. Bohm and L. Foldy, *Phys. Rev.*, **72**, 649 (1947).

none of these classical systems had the particular relation, E = hv. But the essential fact is that they have *some* definite relation between the energy and frequency, which depends on the specific form of the equations of the oscillator, which in turn depends on the specific physical system under consideration. Thus, what is suggested is that we have to deal at the sub-atomic level with some kinds of system that has a non-linear set of equations governing its oscillations, but just such a type of non-linearity as to lead to discrete frequencies of oscillation satisfying the Einstein relation E = hv (and also the de Broglie relation, $p = h/\lambda$). Once this is achieved, then we would automatically be able to explain the appearance of discrete energy levels in matter *and* the appearance of electromagnetic energy in quanta. Moreover, the transitions between discrete energy levels would also be explained, for it is well known from classical systems satisfying non-linear equations that between the stable frequencies of oscillations exist unstable regions, in which the system tends rapidly to move from one stable mode to another. If we suppose that these transitions are very rapid compared with processes taking place at the atomic level, then as far as purely atomic phenomena are concerned they may be regarded as effectively discontinuous. Nevertheless, at a deeper level, they are continuous. Thus, we explain the "atomicity of energy" at a certain level, but can also conceive of the division of these "indivisible atoms of energy" at a more fundamental level; as, for example, atoms of matter, originally found to be indivisible at a certain level, were later found to be analysable into electrons, protons, and neutrons at a more fundamental level.

4. THE INDETERMINACY PRINCIPLE

In general qualitative terms, we see from the preceding discussion that the prospects for finding a causal explanation of the quantum theory were by no means hopeless. Nevertheless, most physicists at the time were very reluctant to embark on such a path, for various reasons, partly of a practical order and partly of a philosophical order. The practical reasons were that to do so, one would have to develop a complex theory on the basis of little experimental evidence, requiring, moreover, the solution of some as yet unsolved mathematical problems of the highest order of difficulty (i.e. the properties of the solutions of non-linear equations). The philosophical reasons were based on the famous indeterminacy principle of Heisenberg, which we shall now proceed to discuss.

To come to the indeterminacy principle, we may ask ourselves the following question: "Suppose that there is some underlying irregular but precisely defined motion of the electron, arising at a deeper level, which could explain the probabilities defined by the Schrödinger wave function, ψ. Would it be possible by actually observing the motion to ascertain its character?" Thus, the first moderately direct determination of the character of random motion at the atomic level came from observations of the related irregular Brownian motions of bodies that were small, but nevertheless much bigger than an atom. Similarly, the character of the random motion at a sub-atomic level would perhaps be indicated by some residual irregularity in the motion of an electron in an atom.

To answer this question, we may analyse the process of observation of an atomic particle, such as an electron, as it goes around the nucleus. Let us first consider the process of observing it with a microscope. Now, because the light which is used in a microscope always comes in the form of discrete packets or quanta, we cannot avoid disturbing the electron when we look at it. For we must use at least one quantum of light to see it. And when this quantum collides with the electron, there will be a minimum disturbance in the latter's motion, which comes from the light that we use in the process of observation.

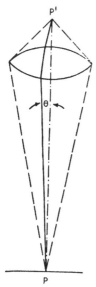

Fig. 4

The energy of a quantum is $E = h\nu$. This suggests that to reduce the disturbance we migh use electromagnetic waves of low frequency, so that we thus obtain a smaller quantum. But here we meet another difficulty. Light not only acts like a particle but also like a wave. It can be demonstrated using the wave theory of light that a light-wave scattered from a definite point, P (see Fig. 4), into a lens, does not form a definite image point, P¹, but instead forms a definite image point, P^1, but instead forms a small poorly defined image region which is proportional in size to the wave-length. But the wave-length is inversely proportional to the frequency, ν. Thus, if the frequency is low, the wave-length will be big; and the image in the microscope will be so poor that we will not know exactly where the electron is.

We are therefore confronted by two difficulties. Because of the particle character of light, we cannot avoid disturbing the particle momentum, creating an unpredictable and uncontrollable disturbance* which we denote by Δp. Because of the wave character of

* The disturbance is unpredictable and uncontrollable because *from the existing quantum theory* there is no way of knowing or controlling beforehand what will be the precise angle, θ, with which the light quantum

light, we cannot avoid an uncertainty, $\Delta\chi$, in the position of the electron, coming from lack of sharpness of the image. A simple calculation which we shall not, however, give here leads to the indeterminacy relations of Heisenberg, $\Delta p \Delta\chi \cong h$, where h is Planck's constant. This result shows that there is a reciprocal relationship between the possible precision of definition of the momentum and that of the position. The more accurately the position is determined, the less accurately the momentum can be defined, and vice versa. This is because an accurate definition of the position requires the use of light of short wave-lengths, so that a large but unpredictable and uncontrollable momentum is transferred to the electron; while an accurate determination of the momentum requires light quanta of very low momentum (and therefore long wave-length), which leads to a large angle of diffraction in the lens and a correspondingly poor definition of the position.

At first sight, one might try to reduce the indeterminacy in the measurement by observing the electron by means other than interaction with light quanta. For example, there have been developed electron microscopes which are able to bring beams of electrons to a focus. Thus, we could substitute a beam of electrons for the beam of light quanta. A more extensive study of the problem shows, however, that the situation would not be changed essentially by this procedure. For, as we have seen, *all* energy comes in quanta, and *all* matter shows the property of acting both like a wave or like a particle. The quantum theory, therefore, implies that the indeterminacy relationships will also apply to observations taken with the electron microscope and indeed to those taken with any other kind of apparatus that we might care to try.

The fact that the quantum theory implies that *every* process of measurement will be subject to the same ultimate limitations on its precision led Heisenberg to regard the indeterminacy relationships, such as $\Delta p \Delta\chi \cong h$, as being a manifestation of a very fundamental and all-pervasive general principle, which operates throughout the whole of natural law. Thus, rather than consider the indeterminacy relationships primarily as a deduction from the quantum theory in its current form, he postulates these relationships directly as a basic law of nature and assumes instead that all other laws will have to be consistent with these relationships.* He is thus effectively supposing

will be scattered into the lens. This leads to an uncertainty in the momentum transfer to the electron. If we knew the angle, θ, we could solve for the momentum transfer ($\Delta p \cong h \sin \theta / \lambda$), and correct for it, so that there would be no uncertainty in the momentum of the electron.

* W. Heisenberg: *The Physical Principles of the Quantum Theory*, Dover Publications (1930), (See. P. 3).

that the indeterminacy relationship should have an absolute and final validity, which will continue indefinitely, even if, as seems rather likely, the current form of the quantum theory should eventually have to be corrected, extended, or even changed in a fundamental and revolutionary way. Indeed, the general point of view of Heisenberg (and of most of the proponents of the usual interpretation of the quantum theory) is that future developments in physics will, if anything, be in the direction of making the behaviour of things even less precisely definable than is possible in terms of the current quantum theory,* so that the current form of the quantum theory sets a limit to the precision of all possible measurements which, in any case, could certainly not be exceeded.

5. RENUNCIATION OF CAUSALITY IN CONNECTION WITH THE ATOMIC DOMAIN A CONSEQUENCE OF THE INDETERMINACY PRINCIPLE

The indeterminacy principle raised a number of important new philosophical questions not appearing in classical mechanics. These questions helped to lead physicists, as we shall see, to renounce the concept of causality in connection with the atomic domain, and thus to adopt a very different philosophical point of view from that which had prevailed up to the advent of the modern quantum theory.

To appreciate the rôle that the indeterminacy principle played in helping to bring about a renunciation of causality, let us recall that in classical mechanics it is just the initial values and initial rates of change of all the mechanical variables defining the state of a given system that determine the future motions of the system in question. According to the indeterminacy principle, however, there exists a fundamental limitation, arising from the very laws of nature at the quantum-mechanical level, which is such that we are unable to obtain the data needed to specify completely the initial values of the various parameters that determine the behaviour of such a mechanical system.

Of course, one might assume that the indeterminacy in the position and the momentum of the electron is a consequence of the fact that these variables, which after all are suggested only by the extrapolation of classical physics to the atomic domain, are not a complete description of the electron. Instead one might suppose that a more nearly complete description would require qualitatively new kinds of variables (such as, for example those discussed in Section 3

* Thus, for example, Heisenberg has recently proposed that for distances shorter than a certain "fundamental length" of the order of 10^{-13} cm., even the properties of space and time should cease to be definable.

in connection with a possible sub-quantum mechanical level), variables not even appearing in the current quantum theory. Hence, if we define only the position and the momentum variables, which provide an adequate approximation to a determinate specification solely at the large-scale level, we will discover that the behaviour of the electron cannot be predicted, because determining factors that are important at the atomic level have been left out of the theory.

The proponents of the usual interpretation of the quantum theory, however, have adopted the hypothesis of Heisenberg, mentioned at the end of Section 4. They do not assume that the indeterminacy principle is just a consequence of the quantum theory in its current stage of development, which could therefore turn out to have only a limited range of validity if, as seems likely it is later discovered that the present form of the theory must be modified, corrected, or extended. Instead they assume that this principle represents an absolute and final limitation on our ability to define the state of things by means of measurement of any kinds that are now possible or that ever will be possible.

If one makes the assumption described above, then one comes to a conclusion having very far-reaching consequences. For even if a sub-quantum mechanical level containing "hidden" variables of the type described previously should exist, these variables would then never play any real rôle in the prediction of any possible kind of experimental results. In fact, if this hypothesis is true, the future behaviour of a system would, at least as far as *we* are concerned, be predictable to just that degree of accuracy corresponding to the limits set by the indeterminacy principle, and to no higher degree. Thus, it is concluded that the present general form of the quantum theory is able to deal with every kind of measurement that we could possibly carry out. Any theory (such as one involving "hidden" variables) which claims to deal with more than this would then be just a metaphysical exercise of the imagination, because nothing in physics would be different from what it would be if these "hidden" variables did not exist.*

---

* For this reason, the term "indeterminacy principle" is more appropriate than the more commonly used term "uncertainty principle". For, as far as any physically observable variables are concerned, it is not to be supposed that these are just "uncertain" to us, because we are not able to measure them with complete precision. Rather, one assumes that their very mode of being requires them to be indeterminate. Of course, the "metaphysical" hidden variables might be uncertain, but since they would in any case never be observable, their being uncertain would have no more real meaning to us than the number of angels that can dance on the head of a pin.

The above conclusion has been made even stronger with the aid of a theorem of von Neumann.* For according to this theorem it would not only be impossible to verify *experimentally* any causal theory that aimed to predict the detailed behaviour of an individual system at the atomic level, but it is impossible even to *conceive* of such an explanation. For von Neumann proved that no conceivable distribution of motions of "hidden" parameters in the observed system could lead to precisely the same results as those of Schrödinger's equation, plus the probability interpretation of the wave function. Thus, for example, one could no longer imagine that even a Laplacian super-being who obtained information without disturbing the system by means of a measurement could make precise predictions about the future. In this way, the indeterminacy principle is supplemented; for the impossibility of making measurements more precise than the limits set by this principle should, according to the theorem of von Neumann, be regarded as a result of the fact that nothing even exists corresponding to a set of "hidden" parameters having a degree of precision of definition going beyond these limits. Thus, the renunciation of causality in the usual interpretation of the quantum theory is not to be regarded as merely the result of our inability to *measure* the precise values of the variables that would enter into the expression of causal laws at the atomic level, but, rather, it should be regarded as a reflection of the fact that no such laws exist.†

We see, then, that the assumption of the indeterminacy principle as an absolute and final law that is supposed to apply to all processes that can possibly take place in the world implies a breakdown of causality in connection with phenomena that depend significantly on the laws of the atomic domain. And in this respect, it must be emphasized that such phenomena are not restricted to just those things that go on in the atomic domain alone, nor just to "hidden" or unobservable properties of things. Real and observable physical phenomena are being assumed to have no causes. For observing apparatus is now available that is sensitive enough to respond in a macroscopically observable way to the properties of *individual* atoms and *individual* quanta of electromagnetic radiation (for example, Geiger counters, Wilson chambers, scintillation counters, etc.). In general, however, measurements carried out on individual atoms or individual quanta with such types of apparatus produce

---

* J. von Neumann, *Mathematische Grundlagen der Quantenmechanik* (Verlag. Julius Springer, Berlin, 1932).

† Thus, the name of "indeterminacy principle" is further justified, for now we are led to conclude that the question of "metaphysical" variables about whose values we are uncertain cannot even arise.

results that show an irregular fluctuation from one observation to the next, with a regular mean behaviour in a statistical aggregate containing a large number of observations. This regular mean behaviour can be predicted to a high degree of approximation in terms of the present quantum theory from the Schrödinger wave function, ψ, with the aid of the probability. But the existing quantum theory yields no way even in principle of predicting how the *individual* measurements will fluctuate from one case to the next. More than this, it does not even have anything in it to which we might at least qualitatively ascribe the origin of any particular individual fluctuation. Of course, as we have already pointed out, we might consider the possibility that such fluctuations originate in irregular motions of some new kinds of entities at a deeper level. But here it is concluded from the indeterminacy principle that even if such a deeper level exists the properties of the new entities could never be measured with sufficient precision to make possible a precise prediction of the irregular fluctuations in the results of individual measurement processes, while von Neumann's theorem implies that such a deeper level of precise causal law could not even exist. Thus, one is led to the conclusion that the precise manner of occurrence of these irregular fluctuations cannot be traced by means of experiments to any kind of causes at all, and that, indeed, it does not even have any causes.

In this respect, the irregular fluctuations treated in the quantum theory are conceived of as being quite different from all other types of irregular fluctuations that have ever been encountered. For example, in Chapter I, we considered the statistical distribution in the rate of automobile accidents, which fluctuates irregularly from day to day, and from place to place. Even the precise details of such fluctuations are found in general, however, to be traceable to a large number of contributing causes, many of which are admittedly very difficult to investigate in detail. Nevertheless, no one doubts, for example, that just what will happen to an individual person in a particular accident (e.g. will a certain bone be broken?, etc.) is determined by suitable causes, some of which may be known and some of which may not be known. But in the usual interpretation of the quantum theory, the precise magnitudes of the irregular fluctuations in the results of individual measurements at the atomic level are not supposed to be determined by any kinds of causes at all, either known or unknown. Instead, it is assumed that in any particular experiment, the *precise* result that will be obtained is *completely arbitrary* in the sense that it has no relationship whatever to anything else that exists in the world or that ever has existed. Thus, we have an example of the conception of "completely lawless" fluctuations discussed in Chapter II, Section 16.*

From the way in which the indeterminacy principle was proved, one might perhaps obtain the impression that the irregular fluctuations in the results of measurements of the properties of individual atoms do, after all, have a cause; for they have been ascribed to the disturbance of the observed object by the measuring apparatus. A more careful analysis shows, however, that one cannot consistently make any *precise* ascription of causes in this way, within the framework of the usual interpretation of the quantum theory. For it must be remembered that the observing apparatus is also subject to the laws of the quantum theory. Thus, to suppose that there exist in the observing apparatus a set of well-defined but irregularly distributed causal factors which in principle determine precisely what will be the disturbance of the observed system in each individual measurement process would, according to the indeterminacy principle, merely be a purely metaphysical assumption, since no additional measurements carried out on the observing apparatus itself could ever determine the precise conditions of these hypothetical causal factors. Moreover, according to von Neumann's theorem, the same impossibility of conceiving of precisely definable causal factors would hold for the observing apparatus as holds for the observed system. Hence, in the usual interpretation of the quantum theory, there is simply no room anywhere for the assumption of additional causal factors to which one could even in principle ascribe the origin of the precise details of the irregular fluctuations in the results of measurements of the properties of individual atoms.

The above conclusion has been made even sharper as a result of an example suggested by Einstein, Rosen, and Podolsky† which gives a case in which one can demonstrate explicitly the inconsistency of supposing that the precise details of the fluctuations described by the indeterminacy principle could be ascribed to disturbances of the

---

* To emphasize what this conception means in practice, consider the process of alpha-particle emission in the radio-active decay of a nucleus, for example, of uranium. In a large aggregate of such nuclei, the precise time of decay of an individual nucleus fluctuates irregularly from one nucleus to another, but the *mean* decay time is predictable, and equal to about two thousand million years. Now consider any one of these individual nuclei, the decay of which can be detected by means of a Geiger counter. Whether this nucleus will decay tomorrow, next week, or in two thousand million years from now is something that the present quantum theory cannot predict. According to the usual interpretation, however, *nothing* determines this time. It is supposed to be completely arbitrary, and not capable of *ever* being related to anything else by means of any kinds of laws at all.

† A. Einstein, N. Rosen, and B. Podolsky, *Phys. Rev.*, **47**, 777 (1935).

observed object by the observing apparatus. In answer to this example, Bohr* pointed out that in the usual interpretation of the quantum theory one must regard the measuring apparatus and observed object as a *single indivisible system*, because they are united by an indivisible quantum which connects them during the process of interaction.† The quantum must belong somehow both to the observed object and to the measuring apparatus, and yet it must be indivisible. This is possible only if the combined system consisting of the observing apparatus and the observed system is, in some sense, a single indivisible entity which cannot correctly be analysed (even conceptually) into more elementary parts. Thus, there could be no meaning to the effort to trace the observed fluctuations in the results of individual measurements to causal factors existing in one part (the measuring apparatus), since it must not even be conceived of as a distinct part, anyway. Hence, one has no choice in the usual interpretation of the quantum theory but to give up altogether the notion that the precise details of the observable fluctuations in the results of individual measurements of a quantum-mechanical level of accuracy have some as yet unknown kinds of causes; and instead, one has to assume that these details are completely arbitrary and lawless.

6. RENUNCIATION OF CONCEPT OF CONTINUITY IN THE ATOMIC DOMAIN

In addition to leading to the conclusion that causality does not apply to the details of individual fluctuation processes connected with the atomic domain, the usual interpretation of the quantum theory has led to a renunciation of the concept of continuity of motion within the same domain.

To show how this has come about, we shall begin by considering certain experimental methods of observing the approximate positions and velocities of an individual electron. Thus, if a free electron of high energy passes through a photographic plate, it leaves a record of its track in the form of small grains of silver, which appear in the microscope more or less as shown in Fig. 5. These grains of silver are deposited as a result of the interaction of the electron with atoms near which it passes: but this interaction must take place in the form of quanta; hence the indeterminacy principle will apply. Now, the appearance of a grain of silver enables us to conclude that an electron passed near enough to permit an interaction. Thus,

* N. Bohr, *Phys. Rev.*, **48**, 696 (1935).

† For a fuller discussion of this problem, see Paul Arthur Schilp, editor, *Albert Einstein, Philosopher Scientist* (Library of Living Philosophers, Evanston, Illinois, 1949).

the grains of silver approximately localize the path of the electron. But because of the indeterminacy principle, we know that an indeterminate momentum was transmitted to the electron in each interaction, so that we cannot predict exactly where the electron will go after it leaves the photographic plate. This indeterminacy is, however, small, so that it is only in a very accurate measurement that it becomes important. But if we needed a very accurate prediction of the notion, we could not be able to get it in this way or in any other way.

According to our customary way of reasoning, we would suppose that the track of grains of silver indicates that a real electron moves continuously through space in a path somewhere near these grains, and by interaction caused the formation of the grains. But according to the usual interpretation of the quantum theory,

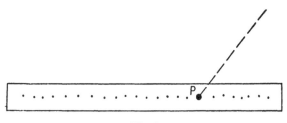

Fig. 5

it would be incorrect to suppose that this really happened. All that we can say is that certain grains appeared, but we must not try to imagine that these grains were produced by a real object moving through space in the way in which we usually think of objects moving through space. For although this idea of a continuously moving object is good enough for an approximate theory, we would discover that it would break down in a very exact theory. Moreover, if we tried to see by experiment whether an electron really moved between the points on the track, for example, by means of a very precise microscopic observation of the position as it passed some point, say P, we would discover that because of the transfer of a quantum the track would change in an unpredictable and uncontrollable way and become another track (as shown in the figure). Thus, according to this view, the notion of a moving electron which supplies a continuous connection between the points at which a track is observed is at best a purely metaphysical one that could never be subjected to experimental verification.

At this point, one might suggest that perhaps the notion of the electron as a small moving body is too simple for this problem, and that our difficulties come from the effort to force our concept of the electron into a mould that was after all suggested mainly by experience in the classical domain. In the quantum domain, it may therefore be necessary to introduce concepts more complex and subtle than that of a particle moving through space in a line and leaving a track. For example, in terms of the notion of a sub-quantum mechanical level discussed in Section 3, it would be possible for the electron to be a very complex structure in a deeper sub-stratum or field, which had a tendency to behave in a wave-like way and yet to produce particle-like concentrations of energy.* The complex entity could leave an approximately localized track of water droplets, which would be produced, for example, by the concentrations of energy described above.

We must emphasize here, however, that the point of view of Bohr and other proponents of the usual interpretations of the quantum theory implies that any explanation of the appearance of the series of droplets in terms of some deeper sub-stratum of continuous motion which leads to concentrations of energy on the water droplets is just as metaphysical as is the explanation in terms of a moving particle. In other words, in a consistent formulation of this point of view, it is assumed that the limits on the divisibility of energy transfers which are characteristic of the current quantum theory must apply, uncontradicted and without approximation, in every domain that will ever be investigated. Thus, the indivisibility of quantum transfers, and with it the need for renouncing the concept of continuity, is regarded as an absolute and final characteristic, which will continue to appear no matter how far physics may progress in the future.

7. RENUNCIATION OF ALL WELL-DEFINED CONCEPTUAL MODELS IN THE MICROSCOPIC DOMAIN—THE PRINCIPLE OF COMPLEMENTARITY

We see from the preceding sections that the usual interpretation of the quantum theory requires us to renounce the concepts of causality and continuity of motion at all possible levels, at least as far as a very precise treatment is concerned. The renunciation implies what Bohr has called an "irrational trait" in the process of transfer of a quantum from one system to another.† By this one means that no rational concept of the details of the process can ever be obtained. This is because if continuity and causality are given up, then one is no longer able to describe or even to think about any well-defined connections between the phenomena at a given time and those at an

---

* A model of this type will be discussed in Chapter IV, Sections 2 and 6.
† N. Bohr: *Atomic Theory and the Description of Nature.*

earlier time. As a result, we have no way to express precisely the qualities and properties that might define the modes of being of individual micro-objects, or to formulate precise laws that might apply to such micro-objects. A similar point of view is indeed already implicit in Bohr's conclusion, described in Section 5, that in an observation carried out to a quantum-mechanical level of accuracy, one must regard the entire ensemble consisting of observing appar-.atus and observed system as indivisibly united by the quanta connecting them. Thus, because the process in which a quantum is transferred from one system to the other cannot be subjected to a detailed rational analysis, there is no way to describe precisely the properties and qualities that characterize the system under observation, as distinguished from those of the observing apparatus.

In order to illustrate the meaning of the point of view described above, let us consider as an example the process of observing an atom with the aid of suitable apparatus. As we have seen, the quanta by which the observing apparatus interacts with the atom will change the latter in a way that cannot be predicted, controlled, described, or even conceived of. Hence, each different apparatus in a sense creates a different kind of atom. Even this terminology is too picturesque, however, because it implies an atom having definite properties when it is not observed, which are changed by interaction with a measuring apparatus. But in the usual interpretation of the quantum theory, *an atom has no properties at all when it is not observed.* Indeed, one may say that its only mode of being is to be observed; for the notion of an atom existing with uniquely definable properties of its own even when it is not interacting with a piece of observing apparatus, is meaningless within the framework of this point of view.

If the notion of the objective existence of atoms and other such micro-objects with uniquely definable properties of their own must be given up, then the question naturally arises, "With what does the quantum mechanics actually deal?" The answer given by Bohr (and within the framework of the usual interpretation the only possible consistent answer) is that it deals not with the properties of micro-objects as such but, rather, *with nothing more than the relationships among the observable large-scale phenomena.* The phenomena are, however, considered as indivisible wholes, which it would be wrong to analyse, even abstractly and conceptually, as made up approximately of different parts, consisting of various kinds of micro-objects.* The rôle of the theory is then regarded as merely the calcu-

* The idea is roughly similar to that of Gestalt psychology, which, roughly speaking, assumes that our perceptions and ideas have properties similar to those ascribed by Bohr to matter in general; that is, they are "wholes" not analysable even abstractly into parts.

lation of the probability distributions for the various possible types of phenomena.

It is clear that the point of view described above leads us to renounce our hitherto successful practice of thinking of an individual system as a unified and precisely definable whole, all of whose aspects are, in a manner of thinking, simultaneously and unambiguously accessible to our conceptual gaze. Such a system of concepts, which is sometimes called a "model", need not be restricted to pictures, but may also include, for example, mathematical concepts, as long as these are supposed to supply a precise representation of the objects that are being described. The usual interpretation requires us, however, to renounce even mathematical models. Thus, the Schrödinger wave function, ψ, is in no sense regarded as a conceptual model of an individual system, since it does not provide a precise description of the behaviour of this system, but permits at most a description of the mean behaviour of a statistical aggregate of systems.

The general point of view described above has been given its most consistent and systematic expression by Bohr, in terms of the principle of complementarity. For a comprehensive treatment of this principle the reader is referred to other sources.* Here we shall give only a brief summary.

In place of a precisely definable conceptual model, the principle of complementarity states that we are restricted to complementary pairs of inherently imprecisely defined concepts, such as position and momentum, wave and particle, etc. The maximum degree of precision of either member of such a pair is reciprocally related to that of the opposite member. The specific experimental conditions then determine how precisely each member of a complementary pair of concepts should be defined in any given case. But no single over-all concept is supposed ever to be possible, which would represent *all* significant aspects of the behaviour of an individual system precisely.

The essential step made by Bohr was then to demonstrate that the laws of the quantum theory permit one consistently to renounce the notion of unique and precisely defined conceptual models in favour of that of complementary pairs of imprecisely defined models. Thus, he was able to prove that the use of complementary pairs of imprecisely defined concepts provides a possible way of discussing the behaviour of matter in the quantum-mechanical domain. But then Bohr's general point of view concerning the principle of comple-

---

* For a very thorough discussion of this principle, see Paul Arthur Schilp, editor, *Albert Einstein, Philosopher Scientist* (Library of Living Philosophers, Evanston, Illinois, 1949).

mentarity goes further than this. For his assumption that the basic properties of matter can *never* be understood rationally in terms of unique and unambiguous models implies that the use of complementary pairs of imprecisely defined concepts will be necessary for the detailed treatment of every domain that will ever be investigated. Thus, the limitations on our concepts implicit in the principle of complementarity are regarded as absolute and final.

8. CRITICISM OF CONCLUSIONS DRAWN IN THE USUAL INTER-
PRETATION OF THE QUANTUM THEORY ON THE BASIS OF THE
INDETERMINACY PRINCIPLE

The conclusions cited in the previous sections have been adopted by most theoretical physicists, who felt that, although they were perhaps difficult to accept, they were in a considerable measure a necessary consequence of the experimental facts leading to the quantum theory. There were, however, a few physicists such as Einstein and Planck, who continued to believe that one should seek a more complete theory that would explain the individual quantum processes in more or less the manner indicated in Section 3. Now one may ask how such a point of view could be maintained, despite the indeterminacy principle which would seem to indicate that at best such a theory would be merely an empty metaphysical speculation that could never be verified by experiment; if not, as suggested by von Neumann's theorem, just simply an impossibility. The answer is that in the chain of reasoning from which these conclusions have been drawn there are serious weaknesses.

Let us begin with a discussion of the indeterminacy principle. We recall that in the proof of this principle, it was essential to use three properties; namely, the quantization of energy and momentum in all interactions, the existence of wave-like and particle-like aspects of these quanta, and the unpredictable and uncontrollable character of certain features of the individual quantum process. It is certainly true that these properties follow from the current general form of the quantum theory. But the question we raised in Section 3 was precisely that of whether or not there exists a deeper sub-quantum mechanical level of continuous and causally determined motion, which could lead to the laws of quantum mechanics as an approximation holding at the atomic level. For if such a sub-quantum mechanical level exists, then, as we have seen, the basic assumptions cited above, which are necessary for the validity of the indeterminacy principle, would not hold at this lower level.* Hence, the indeter-

---

* For example, a quantum could be divided, and in principle predicted and controlled in terms of the new kinds of causal factors existing at this level.

minacy principle simply has nothing whatever to say about the precision that might be obtained in measurements that utilize physical processes taking place at such a sub-quantum mechanical level.

It would appear, therefore, that the conclusions concerning the need to give up the concepts of causality, continuity of motion, and the objective reality of individual micro-objects have been too hasty. For it is quite possible that while the quantum theory, and with it the indeterminacy principle, are valid to a very high degree of approximation in a certain domain, they both cease to have relevance in new domains below that in which the current theory is applicable. Thus, the conclusion that there is no deeper level of causally determined motion is just a piece of circular reasoning, since it will follow only if we assume beforehand that no such level exists.

A rather similar analysis can be made with regard to von Neumann's theorem. For the theorem is based on the implicit assumption that no matter how far we may go in our studies of nature, we shall always find that the state of a system can be defined at least in part with the aid of "observables" which satisfy certain rules of the current quantum theory.* Von Neumann then raised the question, "In addition to these 'observables', are there any other at present 'hidden' variables, which would help define the state of the system more precisely than is now possible in terms of the current formulation of the quantum theory?" His proof that this is impossible depends, in an essential way, however, on the assumption that at least part of the specification of the state of the system will always be in terms of these observables, while the hidden variables will at most serve to make more precise the specification already given by the observables. Such an assumption evidently severely limits the forms of the theories that may be taken into consideration. For it leaves out the important possibility that as we go to a sub-quantum mechanical level the entire scheme of observables satisfying certain rules that are appropriate to the quantum-mechanical level will break down, to be replaced by something very different. In other words, it is quite possible that the whole system of observables applies to a good degree of approximation in the usual quantum-mechanical domain but becomes completely inapplicable in the treatment of deeper lying levels. In this case, the proof of von Neumann's theorem would not be relevant, since the conditions con-

* For example, their eigen values are obtained from linear Hermitean operators, their probability distributions from the square of the absolute value of the coefficient in a suitable expansion of the wave function, etc.

sidered here go beyond the implicit assumptions needed to carry out the proof.*

We see, then, that both in the case of the indeterminacy principle and in that of von Neumann's theorem, conclusions have been drawn concerning the need to renounce causality, continuity, and the objective reality of individual micro-objects, which follow neither from the experimental facts underlying the quantum mechanics nor from the mathematical equations in terms of which the theory is expressed. Rather, they follow from the assumption (usually implicit rather than explicit) that certain features associated with the current formulation of the theory are absolute and final, in the sense that they will never be contradicted in future theories and will never be discovered to be approximations, holding only in some limited domain. Such an assumption so severely limits the possible forms of future theories that it effectively prevents us from considering a sub-quantum mechanical level in which could take place new kinds of motion to which would apply new kinds of causal laws.

We may now ask why the proponents of the usual interpretation have made assumptions of so far-reaching a character on the basis of no experimental or theoretical evidence. The full consideration of this question would require a book in itself, but here we shall content ourselves with giving two of the most important reasons.

First of all there appears to have existed, especially among those physicists such as Heisenberg and others, who first discovered the quantum theory, a rather widespread impression that the human brain is, broadly speaking, able to conceive of only two kinds of things, namely, fields† and particles.‡ The reason generally given for this conclusion is that we can only conceive of what we meet in everyday experience, or at most in experience with things that are in the domain of classical physics where, as is well known, all phenomena fall into one or other of these two classes. Thus, when we come to the microscopic domain, where, as we have seen, neither the field nor the particle concept is adequate, we are supposed to have passed beyond the domain of what we can conceive of. It is found, however, that even in this domain we can predict certain statistical results, with the aid of suitable mathematical calculations. Thus, it

---

* Von Neumann's theorem makes additional assumptions that need not be true. See D. Bohm, *Phys. Rev.*, **85**, 166, 180 (1952).

† Let us recall that fields have usually manifested themselves as waves in physics.

‡ For an exposition of one of the earliest proposals of this point of view, see, for example, W. Heisenberg, *The Physical Principles of the Quantum Theory*, p. 5. A very clear and comprehensive exposition of a similar point of view is also given by C. F. von Weizsacker, *The World View of Physics*, London (1952), p. 104.

is concluded that conceptual thinking will be restricted to the classical domain only, while outside this domain the only thing left to do will be to engage in purely technical manipulations of mathematical symbols according to suitable prescriptions which it is the business of theoretical physicists to discover. As a result, any effort at conceiving of a sub-quantum mechanical level is foredoomed to failure, since even if such a level should actually exist, we could never have direct experience with the entities in it, and could, therefore, never hope to imagine what these entities might be like.

The second reason why modern theoretical physicists have not generally been interested in considering the possibility of a sub-quantum mechanical level has been the widespread adoption of the thesis that we should not postulate the existence of entities which cannot be observed by methods that are *already* available. This thesis stems from a general philosophical point of view containing various branches such as "positivism", "operationalism", "empiricism", and others, which began to attain a widespread popularity among physicists during the twentieth century.* Since we do not yet know how to detect the new entities that might exist in the sub-quantum mechanical level, the point of view described above leads us to refrain from even raising the question as to whether such a level exists. Of course, if future experiments should eventually disclose such entities, then we would naturally start to make theories concerning them, but unless and until this happens, the point of view described above directs us to consider only the kinds of quantities appearing in the current theories.

The answer to these two objections to the sub-quantum mechanical level is quite straightforward.

First of all, the notion that our concepts come only from everyday experience evidently leads to an excessively one-sided point of view on this question. For it is well known that the evolution of our concepts has been due also to *scientific experience*. For example, a great part of our concept of the motion of bodies comes from an imaginative analysis of the experimental and theoretical results of the science of mechanics. In this respect, mathematics in general and the differential calculus in particular have played a key rôle in guiding the development of a clear concept of accelerating motion. It is practically impossible to gain such a clear concept on the basis of everyday experience alone, or indeed even on the basis of labora-

---

* A leading nineteenth-century exponent of the positivist point of view was Mach. Modern positivist philosophers appear to have retreated somewhat from the extreme position of Mach, but the reflection of the position still remains in the philosophical point of view implicitly adopted by a large number of modern theoretical physicists.

tory experience alone, not supplemented by such a deeper imaginative analysis. Thus, one of the most difficult problems that Galileo faced in understanding the laws of falling bodies was just to clarify the concept of acceleration, and to do this, he found it quite essential to use an algebraic expression for the speed of motion. Similarly, most of our concept of a wave comes from a theoretical and experimental study of the interference and propagation of waves in various sciences, such as optics and acoustics, and very little comes from actually watching water waves in everyday experience. In this respect, Huyghen's principle as well as various mathematical treatments of interference and wave propagation constitute an essential part of the modern concept of wave motion.

We see, then, that although the primitive concepts gained in everyday experience may well serve as starting-points for research in physics (and in other sciences), the new laws discovered in this research, both those which take a mathematical form and those which have a more qualitative mode of expression, help lead to a progressive enrichment and refinement of these concepts, until in time they develop into something quite different from what they were when the research started. Now that we are faced with the problem of understanding the new kinds of laws to which our research in the quantum domain has led us, the normal continuation of this procedure would, therefore, be to take the field and particle concepts of classical physics as starting-points, and to modify and enrich them in such a way that they are able to deal with the new combination of wave and particle properties that is implied by the quantum theory. Indeed, a number of concrete efforts in this direction have already been made, and we shall describe them in Chapter IV. Of course, we do not expect the process of development of concepts to stop at this point, either, but rather, as happened in classical physics, we may hope that a series of such modifications and enrichments, each of which helps us attain a better understanding of what is happening in the quantum-mechanical domain, will eventually point the way to revolutionary changes in the whole conceptual structure. Naturally, it is quite possible that we may encounter severe difficulties in such efforts to develop radically new concepts appropriate to the quantum-mechanical domain. Nevertheless, the possibility of such difficulties can hardly be regarded as a good excuse for throwing up our hands in despair before any serious efforts in this direction have been made at all and for asserting that our brains are simply not adequate to the job of imagining what we have not met in everyday experience or in experience in the classical domain.

Let us now discuss the second reason generally given for not considering the possibility of a sub-quantum mechanical level, namely, the positivist principle that we should not postulate the existence of entities that we do not already know how to observe. This principle evidently represents an extraphysical limitation in the possible kinds of theories that we choose to take into consideration. The word "extraphysical" is used here advisedly, since we can in no way deduce, either from the experimental data of physics, or from its mathematical formulation, that it will necessarily remain for ever impossible for us to detect entities which we do not at present know how to observe.

There is no reason why an extraphysical general principle could not conceivably serve as a useful working hypothesis. This particular extraphysical principle cannot, however, be said to be a good working hypothesis. For the history of scientific research is full of examples in which it was very fruitful indeed to assume that certain objects or elements might be real, long before any procedures were known that would permit them to be observed directly. The atomic theory is just such an example. For the possibility of the actual existence of individual atoms was first postulated in order to explain various large-scale regularities, such as the laws of chemical combination, the gas laws, etc. On the other hand, it was of course possible to treat these large-scale regularities directly in terms of macroscopic concepts alone, without the need for the introduction of atoms. Certain nineteenth-century positivists (notably Mach) therefore insisted on purely philosophical grounds that the concept of atoms was meaningless and even "nonsensical" because it was not then possible to observe them as such. Nevertheless, evidence for the existence of individual atoms was eventually discovered by people who took the atomic hypothesis seriously enough to suppose that individual atoms might exist, even though no one had actually observed them. We evidently have here a close analogy to the usual interpretation of the quantum theory, which avoids considering the possibility of a sub-quantum mechanical level, because it cannot be observed by methods that are at present available.

The history of the development of science shows very generally that there are two ways in which scientific progress can be made; first, by the discovery of new facts, which ultimately lead to new kinds of concepts and theories; and secondly, by the explanation of a wide range of existing facts in terms of new concepts and theories, which ultimately lead to new kinds of experiments and thus to the discovery of new facts. In the light of this historical experience, positivism is seen to lead to a one-sided point of view of the possible means of carrying out scientific research. For while it recognizes the importance of the empirical data, positivism flies in the face of the historically demonstrated fact that the proposal of new concepts and theories having certain speculative aspects (e.g. the atomic theory) has quite frequently turned out to be as important in the long run as new empirical discoveries have been.

As an alternative to the positivist procedure of assigning reality only to that which we now know how to observe, we are adopting in this book a point of view introduced in Chapter I, and further developed in Chapter V, which we believe corresponds more closely to the conclusions that can be drawn from general experience in actual scientific research. In this point of view, we assume that the world as a whole is objectively real, and that, as far as we know, it has a precisely describable and analysable structure of unlimited complexity. Thus structure must be understood with the aid of a series of progressively more fundamental, more extensive, and more accurate concepts, which series will furnish, so to speak, a better and better set of views of the infinite structure of objective reality. We should, however, never expect to obtain a complete theory of this structure, because there are almost certainly more elements in it than we can possibly be aware of at any particular stage of scientific development. However, any specified element can in principle ultimately be discovered, but never all of them.

The point of view described above evidently implies that no theory, or feature of any theory, should ever be regarded as absolute and final. Thus, with regard to the current formulation of the quantum theory, we are led to criticize assumptions such as those of Heisenberg and Bohr, that the indeterminacy principle and the restriction to complementary pairs of concepts will persist no matter how far physics may progress into new domains. It should be clear, however, that in making such criticisms, it is not our intention to imply that the quantum theory is not valid or useful in its own domain. On the contrary, the quantum theory is evidently a brilliant attainment of the highest order of importance, a theory whose value it would be absurd to contest. And similarly, Born's probability interpretation of the Schrödinger wave function, Heisenberg's indeterminacy principle, and Bohr's demonstration that in the quantum domain matter shows very general kinds of opposing modes of behaviour under different conditions (e.g. wave and particle), must all be recognized as making extremely important contributions to the expression of the laws of the quantum-mechanical domain. What we wish to stress here is, however, that the brilliant achievements of the quantum mechanics in no way depend on the notion that the features mentioned above (or any other features) of the current theory represent absolute and final limitations on the laws of nature. For all these achievements could equally well be obtained on the basis of the more modest assumption that such features apply within some limited domain and to some limited degree of approximation, the precise extent of which limits remains to be discovered. In this way we avoid the making of arbitrary *a priori* assumptions which evidently could not conceivably be subjected to experimental proof, and we leave the way open for the consideration of basically new kinds of laws that might apply in new domains, laws that cannot be considered if we assume the absolute and final validity of certain features of the theories that are appropriate to the quantum-mechanical domain.

Among the new kinds of laws that one is now permitted to consider if one ceases to assume the absolute and final validity of the indeterminacy principle, a very interesting and suggestive possibility is then that of a sub-quantum mechanical level containing hidden variables. As we saw in Section 3, such a framework of law appears to contain the possibility of explaining, at least qualitatively, the main features of the current quantum theory, as approximations holding in an appropriate domain. Moreover, in the discussion given in this section, we see that no good reasons have been adduced for not considering such theories; and indeed, we shall discuss a number of specific examples of this kind of theory in Chapter IV.

9. THE USUAL INTERPRETATION OF THE QUANTUM THEORY A FORM OF INDETERMINISTIC MECHANISM

The assumption of the absolute and final validity of the indeterminacy principle, which implies that the details of quantum fluctuations have no causes at all, evidently resembles very much that underlying the philosophy of indeterministic mechanism discussed in Chapter II, Section 14. It is, however, in certain aspects more subtle, so that without careful analysis of the problem, one may fail to see just what has been assumed.

In the earlier forms of indeterministic mechanism, it was supposed, either explicitly or implicitly, that the whole universe could be described completely and perfectly in terms of nothing more than certain mathematically definable parameters. These parameters were assumed to undergo arbitrary and lawless fluctuations, the probability distributions of which, however, satisfy a set of purely quantitative laws, which are, in fact, the only kinds of laws that were supposed to be satisfied by anything in the whole world. In the quantum theory, the nearest thing that one might find corresponding to such basic mathematical parameters would be the values of the Schrödinger wave function at every point in space and time, which are, as we saw in Section 2, determined in terms of their initial values for all time in terms of Schrödinger's equation. But, as we have pointed out in Section 7, the Schrödinger wave function is not regarded as corresponding precisely to any property of matter that is supposed to exist. Rather, it is regarded essentially as nothing more than an intermediate and purely mathematical symbol which can be manipulated according to certain prescribed rules in such a way as to permit a correct calculation of the probabilities of various kinds of experimental results.

But now the question arises, "What is the significance of the properties whose probabilities can thus be calculated from the ψ function?" As we have seen in Sections 5 and 7, Bohr has shown that in the usual interpretation of the quantum theory such properties must *not* be regarded as existing objectively in the observed system. There do exist, however, certain things which according to his point of view are admitted to be, to all intents and purposes, objective; viz. the observable large-scale *phenomena*.

Let us review briefly the way in which such phenomena are treated in the usual interpretation of the quantum theory. One can calculate the relationships between these phenomena approximately in terms of the laws of classical mechanics, but, as we have seen in Section 5, in the context of greater precision of experimentation one discovers a random fluctuation in the precise details of these phenomena, not explainable by classical theory. It is then assumed that these fluctuations are completely arbitrary and lawless, having no causes at all. Thus, the theoretical explanation and prediction of these details are supposed to lie for ever outside the scope of what we can hope to accomplish in physics, or in any other science. The subject-matter of physics is then by definition inherently and unavoidably restricted to nothing more than the calculation of the probability distributions of the various possible kinds of phenomena, distributions which are derivable from a certain general physical and mathematical scheme, which grew out of Schrödinger's equation.* In other words, it is supposed that there is nothing in the universe that will not eventually be found to fit into this scheme, the general features of which are thus regarded as absolute and final.

It is clear, then, that we have been led to a point of view that is

---

* This scheme is that of a wave function defined in a configuration space, satisfying a purely linear set of equations, and related to the phenomena through the calculation of the probabilities of various "observables", in terms of the mean values of the associated "operators". This scheme leads to the indeterminacy principle as an inherent and inescapable limitation on the precision with which the basic properties of matter can be defined, described, and even conceived of. Indeed, one may define this mathematical scheme as being precisely the one that is needed if the hypothesis of the absolute and final validity of the indeterminacy principle is to be maintained.

precisely that of indeterministic mechanism. The indeterministic mechanism applies, however, neither to the real micro-objects of the type contemplated in earlier indeterministic mechanist schemes, nor even to a set of purely mathematical parameters, such as appear in Schrödinger's equation. *It applies, rather, only to the observable large-scale phenomena.* Thus, by denying the objective reality of the microscopic domain, and by the associated renunciation of causality and continuity, it becomes possible to save the most essential and characteristic feature of the mechanist position; viz. the assumption that every objective and definable property in the world can be described in terms of nothing more than a set of purely quantitative laws of probability fitting into a certain general physical and mathematical scheme that is absolute and final.

The assumption described above is evidently very similar to that of nineteenth-century physicists who regarded the general physical and mathematical scheme of classical physics as having an absolute and final validity. Indeed, just as classical physicists felt that difficulties, such as those arising from the failure of the Rayleigh-Jeans law, were only "small clouds" soon to be dispelled by some change in the *details* of the deterministic kinds of theories that were then currently held, modern physicists feel that the present crisis in physics* will be resolved by revising the details of the general kinds of probabilistic theories that are now current. What is common to both classical physicists and modern physicists is, therefore, a tendency to assume the absolute and final character of the general features of the most fundamental theory that happens to be available at the time at which they are working. Thus, the usual interpretation of the quantum theory represents, in a certain sense, a rather natural continuation of the mechanistic attitude of classical physicists, suitably adjusted to take into account the fact that the most fundamental theory now available is probabilistic in form, and not deterministic.

* This crisis will be described in Chapter IV.

CHAPTER FOUR

Alternative Interpretations of the Quantum Theory

1. INTRODUCTION

IN the previous chapter, we saw that the usual interpretation of the quantum theory requires us to give up the concepts of causality, continuity, and the objective reality of individual micro-objects, in connection with the quantum-mechanical domain. Instead it leads to a point of view in which physics is said to be inherently and unavoidably restricted, in this domain and below, to the manipulation of mathematical symbols according to suitable techniques that permit, in general, the calculation only of the *probable* behaviour of the phenomena that can be observed in the macroscopic domain. These far-reaching changes in the conceptual structure of physics have been based on the assumption that certain features of the current formulation of the quantum theory, viz. the indeterminacy principle and the appearance of a characteristic set of opposing "complementary" pairs of modes of behaviour (e.g. wave-like and particle-like), are absolute and final features of the laws of nature, which will continue to apply, uncontradicted and without approximation, in every domain that will ever be the subject of physical investigation.

In Section 8 of the previous chapter, we have demonstrated that it is not necessary to make this assumption, and that indeed such an assumption constitutes a dogmatic restriction on the possible forms of future theories. In the present chapter, we shall, however, go further and actually sketch the general outlines of some specific theories which allow us to interpret the quantum mechanics in a new way. These theories permit the representation of quantum-mechanical effects as arising out of an objectively real sub-stratum of continuous motion, existing at a lower level, and satisfying new laws which are such as to lead to those of the current quantum theory as approximations that are good only in what we shall call the quantum-mechanical level.

The new theories serve two principal purposes. First of all, they

104

help to put into a more specific form the criticisms of the usual interpretation of the quantum mechanics made in Chapter III. For by furnishing a concrete example of theories that can be constructed from other points of view, they provide a demonstration of the falsity of the hitherto current impression that we had no choice but to adopt the usual interpretation because any other was thought to be inconceivable. Secondly, and perhaps even more important, these theories may serve as useful starting-points in investigations aimed at the understanding of new domains of phenomena that are not yet very well understood.

In connection with the second point mentioned above, let us recall that there now exists a crisis in physics, arising from the inadequacy of current theories in the treatment of phenomena involving very high energies and very short distances (of the order of 10^{-13} cm. or less). Of course, the proponents of the usual interpretation of the quantum theory are, on the whole, quite aware of this crisis. Nevertheless, as we pointed out in Chapter III, Section 4, their general conclusion has been that the success of probabilistic theories of the type of the current quantum mechanics indicates that in the next domain it is very likely that we shall be led to theories that are, if anything, even more probabilistic than those of the current quantum domain. A more careful consideration of the problem shows that this conclusion carries very little weight. Thus, for example, nineteenth-century physicists could equally well have claimed that the unbroken success of the deterministic laws of classical mechanics in three centuries of applications was a very strong indication that progress into new domains would be very likely to lead to laws that were, if anything, even more deterministic than those that already existed. (In fact, many physicists of the time did hope that the laws of classical statistical mechanics would eventually be deduced completely and perfectly from a deterministic basis.) Thus, it would seem that historical experience should teach us not to make simple extrapolations of previous trends, when we come to the question of what is the degree to which the laws of new domains will show a statistical or a precisely determinate character. Rather it seems clear that one should not decide this question *a priori*, but, instead, one should be ready to try various kinds of laws and to see which kind permits the greatest progress in the understanding of the new domains.

It is in just this spirit described above that we wish the theories discussed in this chapter to be considered. We regard them, not as absolute and final laws which we are laying down from *a priori* considerations, or even as definitive theories of the next level to be treated in physics. Rather, we consider them to be purely provisional proposals with which we are beginning and from which we hope to go forward. Indeed, as we shall see, a considerable amount of progress has already been made since proposals of this kind were first formulated, and the theories have undergone, as is to be expected in normal scientific work in any field, a continual process of enrichment and refinement. We hope that this process will eventually lead to superior theories that are qualitatively different from the ones that served as starting-points, yet perhaps related to them in the same sense that a mature person is related to the child from which he began.

2. GENERAL CONSIDERATIONS ON THE SUB-QUANTUM MECHANICAL LEVEL

Before going on to consider in detail some of the specific theories to which we have referred in Section 1, we shall first make a number of general points concerning the sub-quantum mechanical level, which can be discussed without the aid of such specific theories.

We note, first of all, that if one adopts the hypothesis of a sub-quantum mechanical level containing hidden variables, then, as pointed out in Chapter III, Section 3, we are led to regard the statistical character of the current quantum theory as originating in random fluctuations of new kinds of entities, existing in the lower level. If we consider only those entities which can be defined at the quantum-mechanical level alone, these will be subjected to a genuine indeterminacy in their motions, because determining factors that are important (i.e. the hiden variables) simply cannot be defined in this level. Hence, as in the usual interpretation of the quantum theory, we regard the indeterminacy implied by Heisenberg's principle as an objective necessity and not just as a consequence of a simple lack of knowledge on our part concerning some hypothetical "true" states of the quantum-mechanical variables. Thus, it is not the existence of indetermination and the need for a statistical theory that distinguishes our point of view from the usual one. For these features are common to both points of view. The key difference is that we regard this particular kind of indeterminacy and the need for this particular kind of statistical treatment as something that exists only within the context of the quantum-mechanical level, so that by broadening the context we may diminish the indeterminacy below the limits set by Heisenberg's principle.

To go beyond the limits set by Heisenberg's principle, it will be necessary to use new kinds of physical processes that depend significantly on the details of what is happening at the sub-quantum mechanical level. As we shall see later, there is some reason to believe that such processes could perhaps be found in the domain of very high energies and of very short distances. It is clear, however, that

The Sub-Quantum Mechanical Level

in any process which can be treated to an adequate degree of approximation by the laws of the current quantum theory, the entities existing in the lower level cannot be playing any very significant rôle. Very little information about these entities could then be obtained by observing the results of this kind of process. In such an observation, Heisenberg's principle would, therefore, apply to a very high degree of approximation as a correct limitation on how well the state of an individual physical system could be determined, while evidently, if we observed the system with the aid of physical processes sensitive to the precise states of the hidden variables, this limitation would cease to be applicable.

To illustrate in more detail what the indeterminacy principle would mean in terms of a sub-quantum mechanical level, it will be helpful to return here to the analogy of Brownian movement, already considered in Chapter III, Section 3.

As we have seen, the motion of a smoke particle is subject to random fluctuations, originating in collisions with the atoms which exist at a lower level. As a result of these collisions, its motions cannot be completely determined by any variables (e.g. the position and velocity of the particle) existing at the level of the Brownian motion itself. Indeed, the lack of determination is not only qualitatively analogous to that obtained in the quantum theory, but, as has been shown by Furth,[1]* the analogy even extends to the quantitative form of the indeterminacy relations. Thus if we observe a moving smoke particle throughout some short interval of time, Δt, we will find random fluctuations of magnitude ΔX in the mean position, and of magnitude, ΔP, in its mean momentum, which satisfy the relationship†

$$\Delta P \Delta X \cong C$$

Here C is a constant, which depends on the temperature of the gas, as well as on other properties such as its viscosity. If the reader will refer to Chapter III, Section 4, he will see that the form of this relationship is just the same as that of Heisenberg, except that the Planck's constant, h, has been replaced by the constant, C, which depends on the state of the gas.

* The reference numbers refer to the Bibliography at the end of this chapter.

† Basically, this relationship comes from the formula $(\Delta x)^2 = a\Delta t$ for the mean square of the distance moved by the particle in its random motions during the time, Δt. Thus we have for the root mean square fluctuation in the momentum (assuming zero mean velocity to simplify the argument) $\left[\overline{\left(\frac{\Delta x^2}{\Delta t}\right)}\right]^{\frac{1}{2}} = a^{\frac{1}{2}}(\Delta t)^{-\frac{1}{2}}$.

Then, with $\Delta X = [(\Delta x)^2]^{\frac{1}{2}}$, we get $\Delta X \Delta P = ma = C$.

107

There is, however, an important respect in which the analogy between the Brownian motion and the quantum theory is not complete. This difference arises essentially in the fact that C is not a universal constant whereas h is. As a result, in principle at least, one is able by changing conditions suitably to make C arbitrarily small (e.g. by lowering the temperature) and thus reduce the indeterminacy below any desired value. On the other hand, the constant, h, does not depend on conditions in any known way, so that Heisenberg's relations imply, as far as we have been able to tell, an indeterminacy that is universal, at least within the quantum-mechanical domain. This means that while we can by a suitable choice of conditions construct apparatus (e.g. a microscope) which is not significantly affected by the kind of Brownian motion that we wish to observe, we cannot obtain a similar result in connection with the quantum-mechanical indeterminacy. To improve the analogy, we should therefore have to suppose that in the quantum domain we are effectively restricted to using apparatus that is itself undergoing the Brownian motion to an extent that is comparable with that undergone by the micro-systems that we are trying to observe. If we recall, however, that in our point of view, *all* matter in all its known manifestations is continually undergoing fluctuations originating in the sub-quantum mechanical level, we can see that the above extension of the analogy is justifiable. Considering that these fluctuations are present everywhere with essentially the same characteristics, we conclude that the universal and uniform character of the limitations implied by Heisenberg's principle in the quantum domain would not be an unexpected consequence of our assumptions.

To overcome these limitations we should, as we have already pointed out, have to take advantage of properties of matter that depended significantly on the sub-quantum mechanical level. One way to do this would be to make our observations with the aid of processes that were very fast compared with the sub-quantum mechanical fluctuations, so that the whole measurement would be over before these fluctuations could have changed things by very much (just as to photograph a rapidly moving object we need a very fast camera). Such rapid processes are most likely to be obtained in the high energy domain since, from the Einstein relation, $E = h\nu$, a high energy, E, implies a process of high frequency, ν.

Finally, the analogy of Brownian motion also serves to bring out two different limiting modes in which the indeterminacy originating in random sub-quantum mechanical fluctuations may manifest itself. For let us consider, not the Brownian motion of smoke particles, but rather that of very fine droplets of mist. It is evident that there is a certain indeterminacy in the motion of these droplets that could be removed only by going to a broader context, including the air molecules with which these droplets are continually being struck. It remains true, however, that in their irregular Brownian motions the droplets retain their characteristic mode of existence as very small bodies of water. On the other hand, as we approach the critical temperature and pressure of the gas* a new behaviour appears; for the fine droplets begin to become unstable. The substance then enters a phase in which the droplets are always forming and dissolving and, as a result, the substance becomes opalescent.

Here we have a new kind of fluctuation, which leads to an indeterminacy in the very mode of existence of the substance (i.e. between existence in the form of droplets and existence in the form of a homogeneous gas).

Similarly, it is possible that the very mode of existence of the electron will eventually be found to be indeterminate, when we have understood the detailed character of quantum fluctuations. Indeed, the fact that the electron shows a characteristic wave particle duality in its behaviour would suggest that perhaps this second kind of indeterminacy will turn out to be the relevant one; for if such an indeterminacy exists, it would lead to a concept of the electron as an entity that was continually fluctuating from wave-like to particle-like character, and thus capable of demonstrating both modes of behaviour, each of which would, however, be emphasized differently in the different kinds of environment supplied, for example, by different arrangements of laboratory apparatus.

Of course, we have no way at present to decide which of these interpretations of the indeterminacy principle is the correct one. Such a decision will be possible only when we shall have found an adequate theory that goes below the level of the quantum theory. Meanwhile, however, it is important to keep both possibilities in mind. In the subsequent work, we shall therefore discuss examples of both kinds of theories.

3. BRIEF HISTORICAL SURVEY OF PROPOSALS FOR ALTERNATIVE INTERPRETATIONS OF THE QUANTUM THEORY

It is significant to note that the first steps towards an alternative interpretation of the quantum theory were taken about thirty years

---

* The critical temperature and pressure define a point at which the distinction between gas and liquid disappears. Above this point there is no sharp qualitative transition between liquid and gas, while below it such a transformation can take place. If we heat a liquid confined in a strong container past its critical point, the meniscus separating gaseous and liquid phases disappears, showing that there is now only one phase, which may be thought of as a very dense gas.

ago by de Broglie[2] and by Madelung[3] at more or less the same time as the usual interpretation was being brought into its current definitive form. Neither of these steps was, however, carried far enough to demonstrate the possibility of a consistent treatment of all the essential aspects of the quantum theory. Indeed, the interpretation of de Broglie was subjected to severe criticisms by some of the proponents of the usual interpretation.[4] Partly as a result of these criticisms and partly as a result of additional criticisms which he made himself, de Broglie gave up his proposals for a long time.[5]

After these efforts had died out, it was not until about 1950 that a systematic tendency to question the usual interpretation of the quantum theory began to develop on an appreciable scale. Among the most thoroughgoing of the earlier critical efforts in this direction were those of Blokhinzev and Terletzky[6]. These physicists made it clear that it is not necessary to adopt the interpretation of Bohr and Heisenberg, and showed that instead, one may consistently regard the current quantum theory as an essentially statistical treatment, which would eventually be supplemented by a more detailed theory permitting a more nearly complete treatment of the behaviour of the individual system. They did not, however, actually propose any specific theories or models for the treatment of the individual systems. Then in 1951, partly as a result of the stimulus of discussions with Dr. Einstein, the author began to seek such a model; and indeed shortly thereafter[7] he found a simple causal explanation of the quantum mechanics which, as he later learned, had already been proposed by de Broglie in 1927. Meanwhile, however, the theory had been carried far enough so that the fundamental objections that had been raised against the suggestions of de Broglie had been answered. This was done mainly with the aid of a theory of measurements[8] which showed that the new interpretation was consistent with all the essential characteristics of the quantum theory. Partly as a result of this work, and partly as a result of additional suggestions made by Vigier[9], de Broglie[5] then returned to his original proposals, since he now felt that the decisive objections against them had been answered.

At this stage, as pointed out in Section 1, the author's principal purpose had not been to propose a definitive new theory, but was rather mainly to show, with the aid of a concrete example, that alternative interpretations of the quantum theory were in fact possible. Indeed, the theory in its original form, although completely consistent in a logical way, had many aspects which seemed quite artificial and unsatisfactory.[10] Nevertheless, as artificial as some of these aspects were, it did seem that the theory could serve as a useful starting-point for further developments, which it was hoped could modify and enrich it sufficiently to remove these un-satisfactory features. Such developments, which have in fact occurred[11, 12], at least in part, and which are still going on, will be discussed in more detail in Section 5. Meanwhile, however, a number of largely independent efforts have been made in the same general direction by Vigier[9], Takabayasi[13], Fenyes[14], Weizel[15] and many others. While none of the efforts cited above has been able to avoid completely some kinds of artificial or otherwise unsatis-factory features, each of them introduces new ideas that are well worth further study. It is clear, then, that even if none of the alter-native interpretations of the quantum theory that have been pro-posed thus far has led to a new theory that could be regarded as definitive, the effort to find such theories is nevertheless becoming a subject of research on the part of more and more physicists, who are apparently no longer completely satisfied with continuing on the lines of research that are accessible within the framework of the usual interpretation.

4. A SPECIFIC EXAMPLE OF AN ALTERNATIVE INTERPRETATION OF THE QUANTUM THEORY

In this section, we shall sketch in a qualitative way a specific example of an alternative interpretation of the quantum theory. This example is not the original one proposed by the author, but already contains a number of modifications and new features, which are aimed at removing some of the unsatisfactory aspects of the earlier proposals.

We begin by recalling that in the quantum-mechanical domain matter is able, under different conditions, to show either wave-like or particle-like behaviour, so that it is evident that the wave concept and the particle concept are each, *by themselves*, incapable of dealing with the full richness of properties demonstrated by matter in this domain. Now, the first and simplest idea to suggest itself in the face of this problem is that perhaps the difficulty arises out of the fact that in previously existing theories only two possibilities were con-sidered, namely, that of the pure wave and that of the pure particle, these two possibilities being regarded as mutually exclusive. On the other hand, it is evidently possible that in any given process, both wave and particle could be present *together* in some kind of inter-connection. Of course, this proposal does not constitute a very great enrichment of the concepts that were hitherto used, but, as we shall see, it is already able to represent the essential properties of matter in the quantum domain.

We now formulate this point of view in more detail. We first postulate that connected with each of the "fundamental" particles of physics (e.g. an electron) is a body existing in a small region of space. The probable size of this region we shall discuss later, but for the present we assume only that it is smaller than the size of an atom, and indeed so small that in most applications at the atomic level the body can be approximated as a mathematical point (just as in the earliest forms of the atomic theory one was able for many purposes to approximate atoms as points).

The next step is to assume that associated with this body there is a wave, without which the body is never found. This wave will be assumed to be an oscillation in a new kind of field, which is represented mathematically by the ψ field of Schrödinger. In other words, we no longer suppose that the Schrödinger wave function is nothing more than a mathematical symbol convenient to manipulate in order to calculate certain probabilities, but, instead, represents an objectively real field, somewhat like the gravitational and the electromagnetic, but having some new characteristics of its own. Instead of satisfying Maxwell's equations or the equations of the gravitational field, this field satisfies Schrödinger's equation, which provides, however, as in the case of the other fields, a partial differential equation determining the future changes of the field in terms of its value at all points in space at a given instant of time.*

We now assume that the ψ field and the body are interconnected in the sense that the ψ field exerts a new kind of "quantum-mechanical" force on the body, a force that first begins to manifest itself strongly in the atomic domain, so that we can understand why it has not previously turned up in the study of the large-scale domain. We also suppose that the body may exert a reciprocal influence on the ψ field, but that this reciprocal influence is small enough to be neglected in the quantum-mechanical domain, even though it is, as we shall see later, likely to be significant in the sub-quantum mechanical domain.

As to the precise character of the quantum-mechanical force exerted by the ψ field on the body, this is not very important at the level of the theory at which we are working, because a very wide range of kinds of forces could lead to essentially the same results. All that is important for the present is to suppose that the force is such as to produce a tendency to pull the body into regions where $|\psi|$ is largest.†

* The ψ field is complex, but this creates no real difficulty, since we can always write it as $U + iV$ where U and V are real. Thus, the ψ function is just a short-hand way of talking of two coupled real fields (see D. Bohm, *Quantum Theory*, Chapter III).

† Note that the "quantum-force" in this model is quite different from what it was in some of the earlier models, discussed in references (7) and (8). In these earlier models this force was assumed to be derived from a "quantum-potential"—$\hbar^2 \Delta^2 R/2mR$, where $\psi = R e^{iS/\hbar}$ and R and S are

A Specific Example

If the above tendency were all that were present, the body would eventually find itself at the place where the ψ field had the highest intensity. We now further assume that this tendency is resisted by random motions undergone by the body, motions which are analogous to the Brownian movement. These random motions clearly could have many sources. They could, for example, come from random fluctuations in the ψ field itself. Indeed, it has been characteristic of all other fields known thus far that typical solutions to the field equations represent in general only some kind of average motion. For example, real electromagnetic fields do not oscillate in some simple and regular way, but in general they have complicated and irregular fluctuations (e.g. those representing the thermal radiation coming from atoms in the walls of containers, etc.). Similarly, hydrodynamic fields, representing the velocity and density distributions of real fluids, generally show turbulent fluctuations, about an average satisfying certain kinds of simplified hydro-dynamical equations. Hence, it is not unreasonable to suppose that the ψ field is undergoing random fluctuation about an average that satisfies Schrödinger's equation and that these fluctuations communicate themselves to the body. The details of these fluctuations would then represent properties of the field associated with a sub-quantum mechanical level, since the quantum-mechanical level is treated in terms of the mean part, which satisfies Schrödinger's equation. On the other hand, the bodies could obtain a random motion from a sub-quantum mechanical level in other ways, for example, as in ordinary Brownian motion, by direct interaction with new kinds of entities existing in this lower level. Indeed, at the present stage of the theory, it is not relevant where such fluctuations come from. All that is important here is to assume that they exist and to see their effects. The question of their origin can then appropriately be raised only in a study of the sub-quantum mechanical level.

Once admitting the existence of these fluctuations, we then see that they will produce a tendency for the body to wander in a more or less random way over the whole space accessible to it. But this tendency is opposed by the "quantum-force" which pulls the body into the places where the ψ field is most intense. The net result will be to produce a mean distribution in a statistical ensemble of bodies, which favours the regions where the ψ field is most intense, but which still leaves some chance for a typical body to spend some time in the places where the ψ field is relatively weak. Indeed, a rather similar

real. Here no such a specific assumption is needed. The present model has the advantage that it is conceptually simpler than some of the earlier models. Moreover, as we shall see in Section 6, it is closer to what is suggested by efforts to go to the theory of relativity and of electron spin.

behaviour is obtained in classical Brownian motion of a particle in a gravitational field, where the random motion which tends to carry the particle into all parts of the containers is opposed by the gravitational field, which tends to pull it towards the bottom. In this case, the net effect is to produce a probability distribution,* $P = e^{-mgz}/\varkappa T$, which describes a tendency for the particles to concentrate at the bottom and yet occasionally in their random motions to be thrown up to great heights. In the quantum-mechanical problem, one can show by means of a treatment that is given elsewhere† that with physically reasonable assumptions concerning the quantum force and the random motions coming from the sub-quantum mechanical level, we obtain Born's probability distribution, $P = |\psi|^2$.

What is the meaning of this result? It means that instead of starting from Born's probability distribution as an absolute and final and unexplainable property of matter, we have shown how this property

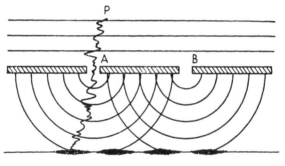

Fig. 6A.

could come out of random motions originating in a sub-quantum mechanical level.

A more detailed treatment (appearing in references (7), (8), and (11)) shows that the above result is sufficient to lead to an interpretation that is consistent with all the essential results of the quantum theory. Here, however, we shall illustrate only one way in which this happens, namely, the explanation of the wave-particle duality. To do this, we consider an experiment in which electrons are sent separately and independently with perpendicular incidence into a system containing two slits, illustrated in Fig. 6A. Every electron is assumed to have initially the same momentum, and

---

* Here m is the mass of the particle, z its height, T is the temperature of the medium, g the gravitational constant, and $\varkappa$ is Boltzman's constant.
† See reference (11).

therefore the same wave function,* which in fact takes the form of a plane wave incident perpendicularly on the slit system. These waves will be diffracted through the slit system as shown in the figure, and a pattern of high and low intensity will be obtained at the detecting screens, just as in the case of light quanta, discussed in Chapter III.

The small body connected with the electron undergoes, however, a random motion. Thus, it follows an irregular path starting out from point P, as indicated in Fig. 6A. Each electron then arrives at the screen at a certain point. After a large number of electrons have passed through the slit system, we will obtain a statistical pattern of such points, in which the density of electrons is proportional to the field intensity, $|\psi|^2$, at the screen. The statistical tendency to appear where $|\psi|^2$ is greatest is due to the effects of the "quantum-force" while the random motions explain why the precise points at which the various particles appear fluctuate in an irregular way.

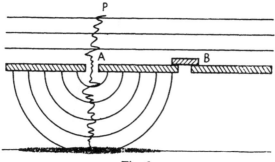

Fig. 6B.

Now suppose that we close slit B. The wave pattern will now, as shown in Fig. 6B, cease to have strong and weak fringes. Thus, a new pattern of electrons is obtained at the screen. Hence, the closing of slit B influences even those particles that pass through slit A, because it influences the "quantum-force" felt by the particle as it moves between the slit system and the screen.

In this way, we can understand how the wave-particle duality originates. On the other hand, in the usual interpretation, no such an understanding is possible. All that we can do is to accept without further discussion the fact that electrons enter the slit system, and

* This follows from the de Broglie relationship, $p = h/\lambda$; for the Schrödinger fields of all the electrons will then take the same form, $e^{ipx/\hbar}$. Actually of course, all waves must consist of packets, but in this case, the packet is so much bigger than the slits that we can approximate it as an infinite plane wave. See, for example, D. Bohm, *Quantum Theory*, Chapter 3.

appear at the screen with an interference pattern. As to how this came about, such a question cannot even be raised within the framework of the usual interpretation. Thus, according to the principle of complementarity, described in Chapter III, Section 7, we may use the wave model to discuss how the slit system governs the statistical interference pattern and we may use the particle model to discuss the fact that individual spots are received at the screen and not a continuous distribution of matter. But no over-all notion is supposed to be available that could even permit us to consider the question of how a single material system, conceived of as following some in principle precisely definable and unambiguous (i.e. unique) course of motion, could bring about *both* the statistical interference pattern *and* the appearance of a series of discrete spots on the screen. But, as we have seen here, this result was achieved by the simple expedient of considering that the electron is a combination of particle and field, interconnected and undergoing suitable random fluctuations in its motions.

Finally, let us note that in our model we have not insisted on a purely causal theory, for we have also utilized the assumption of random fluctuations originating at a deeper level. The essential point here is, however, that the laws of the sub-quantum mechanical level, both causal and statistical, are qualitatively different from those of the quantum level, and lead to those of the latter level only as an approximation. Thus, there is no reason why Schrödinger's equation should have any relevance at the lower level, since this equation was assumed to be an average holding only at the higher level. Indeed, the ψ field itself may well be only an average of new kinds of variables that are defined at the lower level.

It is also quite possible that the approximation in which we neglect the reciprocal action of the body on the ψ field would break down here. Moreover, it is quite evident that the approximation in which we regard the body as a point (and thus neglect its inner structure) must break down in processes taking place at the lower level.

In general, then, one sees that an enormous range of possibilities for new kinds of theories has been opened up, leading to the current theory in the quantum domain as a good approximation, and to extremely different kinds of theories in new domains. These possibilities could not be considered if we maintained the assumption of the absolute and final validity of the indeterminacy principle, and with it the usual interpretation of the quantum theory. And as we shall see later, there is good reason to suppose that some of these possibilities are likely to be helpful in the treatment of the new domains of phenomena associated with very high energies and very short distances.

5. CRITICISMS OF SUGGESTED NEW INTERPRETATION
OF THE QUANTUM THEORY

We shall now discuss a number of significant criticisms of the interpretation of the quantum theory that has been introduced in Section 4.

First of all, it must be pointed out that, as presented thus far, the theory does not take into account certain important problems such as those raised by the electron spin and the theory of relativity. While it is true that Schrödinger's equation, which neglects spin and relativity, is, up to a certain point, a fairly good approximation, it is not adequate either in the domain of very high energies or even in a very precise treatment of the low energy domain. Since it is our express purpose to apply the new interpretation in the domain of high energies, it is evidently necessary that we go on to the consideration of spin and relativity. This will require that we treat the Dirac equation, which takes into account the effects of these general properties.

Secondly, even in the domain of low energies, a serious problem confronts us when we extend the theory given in Section 4 to the treatment of more than one electron. This difficulty arises in the circumstance that, for this case, Schrödinger's equation (and also Dirac's equation) do not describe a wave in ordinary three-dimensional space, but instead they describe a wave in an abstract 3N-dimensional space, when N is the number of particles.* While our theory can be extended formally in a logically consistent way by introducing the concept of a wave in a 3N-dimensional space,† it is evident that this procedure is not really acceptable in a physical theory, and should at least be regarded as an artifice that one uses provisionally until one obtains a better theory in which everything is expressed once more in ordinary three-dimensional space.

Finally, our model in which wave and particle are regarded as basically different entities, which interact in a way that is not essential to their modes of being, does not seem very plausible. The fact that wave and particle are never found separately suggests instead that they are both different aspects of some fundamentally new kind of entity which is likely to be quite different from a simple wave or a simple particle, but which leads to these two limiting

---

* The notion of a 3N-dimensional space is a purely mathematical concept. A point in three-dimensional space can be described by three numbers, which are its three co-ordinates. One can in a purely mathematical way generalize this idea. Thus, with four numbers we can describe the co-ordinates of a point in a four-dimensional space, and with N such numbers the co-ordinates in an N-dimensional space.

† This is done in references (7) and (8).

manifestations as approximations that are valid under appropriate conditions.

It must be emphasized, however, that these criticisms are in no way directed at the logical consistency of the model, or at its ability to explain the essential characteristics of the quantum domain. Rather, they are based on broad criteria, which suggest that many features of the model are implausible and, more generally, that the interpretation proposed in Section 4 does not go deep enough. Thus, what seems most likely is that this interpretation is a rather schematic one which simplifies what is basically a very complex process by representing it in terms of the concepts of waves and particles in interaction.

6. FURTHER DEVELOPMENTS OF THE THEORY

We shall now consider a number of modifications and enrichments of the theory, which are aimed at resolving the problems raised by the criticisms discussed in Section 5.

We begin with the consideration of the problem of spin. Here our first step was to try to extend the theory to the Pauli equation, which takes spin into account, but still neglects the effects of relativity, so that it is good only at velocities that are low compared with that of light.

In order to deal with this problem, we have been led to consider several comparatively new ideas. Our first new step is to cease to approximate the body connected with the electron as a mathematical point. The most elementary new mode of motion that a body of non-zero size can have (relative to that possessed by a point) is rotation about its centre of mass. Indeed, as is shown in any elementary text on mechanics, if such a body is regarded as completely rigid, then its internal motions can be described completely in terms of three angles (the Euler angles) which determine its orientation in space. Of course, no real body can be perfectly rigid, but we may assume that the body with which we have to deal is rigid enough, so that we can, at least as far as the level of interest is concerned, ignore the effects of its lack of rigidity. The electron spin is then interpreted in terms of the rotation of this body, which gives rise to an "intrinsic" angular momentum, over and above that which is due to its orbital motion.

On the basis of this more extensive model, we are then able to arrive at a complete and consistent interpretation of the Pauli equation. We do not wish here to enter into its details, which are given elsewhere,[12] but we shall merely point out that to deal with electron spin it was sufficient to make an extension of the theory, which was quite natural, and which was indeed in a sense even contained implicitly in the previous theory, namely, to take into account the fact that the bodies with which we are concerned are not points.

The next step was to extend this theory to the Dirac equation, and thus to take relativity into account.* We shall not discuss the details of this theory here, for they are purely technical. It is worth mentioning, however, that something new comes out of the model, for we now find that this model permits the ψ field to oscillate in several different ways. In one of these ways, it satisfies the Dirac equation, but as in the case of the model of the Schrödinger equation discussed in Section 4, it does this only in an approximation applying to small oscillations of the field around a randomly fluctuating background that averages out to zero. But then it can oscillate in such a way that certain functions of the ψ field satisfy Maxwell's equation, which, as we saw in Chapter II, Section 6, are the equations satisfied by the electromagnetic field. Moreover, it is found that in a higher approximation these two oscillations are coupled, in the proper way (i.e. in the way that is found to be needed in current theories to lead to correct treatments of the relationship between the electron and the electromagnetic fields). Thus, we are able from a single theory to obtain a unified treatment of two kinds of fields, which were previously simply postulated separately, along with their coupling. Moreover, the theory is rich enough so that it could lead to still more modes of oscillation, and we shall see in Section 8 that these might be important in connection with new kinds of particles, such as mesons, which are found in very high energy processes.

With regard to the second criticism mentioned in Section 5, namely, the need to introduce fields in a multi-dimensional space in order to treat the many-body problem, work now in progress has already gone a long way towards indicating a possible solution to this difficulty.† This work is based on using as a starting-point, not the many-body Schrödinger equation, which is defined in a multi-dimensional space, but rather the so-called "second-quantized" field theory whose basic quantities are defined in a three-dimensional space. This theory is now generally regarded by most theoretical physicists as the best and most fundamental existing formulation of the quantum theory itself.

In this theory the starting-point is to suppose that the basic entities are fields, such as electromagnetic, electronic, mesonic, etc. These fields are then regarded as mechanical systems, subject

---

* This work has only recently been completed and will be published later.

† De Broglie and Vigier have proposed another direction in which a solution to this problem can be sought. See references (5) and (9).

to the general laws of the quantum theory. These laws imply a number of important properties of the fields. These are:

(1) Even in the vacuum, the fields are undergoing violent and very rapid random fluctuations. These fluctuations serve, however, as a uniform background that is not directly observable at the macroscopic level, because they average out to produce a negligible effect at this level.

(2) On top of these random fluctuations there are comparatively small systematic oscillations. These oscillations do not cancel out at the macroscopic level, but add up to produce cumulative effects that are detected there. Matter as it is commonly met at higher levels is the result of these systematic oscillations. Thus, to have an electron in a certain region of space means, in this theory, to have in this region a systematic but localized oscillation, responsible for all the manifestations here which define the properties of the electron (charge, mass, momentum, angular momentum, etc.).

(3) The laws of the quantum theory imply that certain properties of the field will be discrete (e.g. charge, mass, energy, momentum, angular momentum). This discreteness is responsible for the particle-like properties of the field.

In the usual interpretation, the quantum field theory is like all other forms of the quantum theory, regarded as nothing more than a means of manipulating mathematical symbols so as to get correct answers for certain experimental results. Thus, the properties described above are not taken very seriously, and are indeed regarded as at best convenient ways of talking about these mathematical manipulations. On the other hand, we are adopting here a point of view in which we suppose that microscopic processes really take place, and that it is our objective to understand how this happens. Thus, we are led to try to develop further the model described above, which is very strongly suggested by the quantum field theory.

In this connection it is rather interesting that our model of the Dirac equation, obtained on a rather different basis, has many features similar to that suggested by second quantization. Indeed, when the two models are brought together, we are led to a new theory, in which both the Dirac equation and the second quantization theory come out as approximations holding only in the quantum-mechanical level, but not in the sub-quantum mechanical level. It has not yet been possible to work out the full implications of this theory, but already there is much evidence in favour of the following general picture, namely, that the field acts as a wave, and yet (because of non-linear terms in the equations) shows a tendency to produce discrete and particle-like concentrations of energy, charge, momentum, mass, etc. Thus, we are led to a point of view rather like that suggested in Section 2 in connection with the Brownian motion of mist droplets near the critical point, namely, that the particle-like concentrations are always forming and dissolving. Of course, if a particle in a certain place dissolves, it is very likely to re-form nearby. Thus, on the large-scale level, the particle-like manifestation remains in a small region of space, following a fairly well-defined track, etc. On the other hand, at a lower level, the particle does not move as a permanently existing entity, but is formed in a random way by suitable concentrations of the field energy.

Moreover, it is clear that the completion of this model would answer not only the second criticism in Section 5, but also the third one. For in this case, wave and particle aspects of matter would arise out of the motions of a more complex kind of entity existing at a lower level and would not be simply postulated as separate but interacting entities.

It is clear, then, that while we have not yet produced a definitive theory of the sub-quantum mechanical level that would lead to all the features of the current quantum theory in a natural way as approximations holding at a certain level, several lines of research in this direction are now open, so that the prospects for achieving such a theory are by no means distant.

7. THE CURRENT CRISIS IN MICROSCOPIC PHYSICS

We shall now discuss briefly the current crisis in microscopic physics, in order to lay a foundation for an explanation of some of the advantages of the point of view towards the quantum theory that we have developed in this chapter.

Let us first discuss the theoretical aspects of this crisis.* When one applies the existing quantum theory to the electrodynamics of "elementary" particles (such as electrons, protons, etc.), internal inconsistencies seem to arise in the theory. These inconsistencies are connected with the prediction of infinite values for various physical properties, such as the mass and the charge of the electron. All these infinities arise from the extrapolation of the current theory to distances that are unlimitedly small. Among the things that make such an extrapolation necessary, one of the most important is the assumption, which seems to be an intrinsic part of current theories that "elementary" particles, such as electrons, are mathematical points in the sense that they occupy no space at all. On the other hand, in spite of many years of active searching on the part of theoretical

---

* This problem is, in its details, very complex, and cannot be discussed without the aid of a lengthy and elaborate mathematical treatment. We shall, therefore, give here only a qualitative summary of the essential aspects of the problem.

physicists throughout the whole world, no way has yet been found to incorporate consistently into the current quantum theory the assumption that the electron occupies a finite region of space.* While it has been suggested that perhaps the infinities come from an inadequate technique of solving the equations (i.e. perturbation theory), persistent efforts to improve this technique have not yet produced any favourable results, and indeed those results that have been obtained favour the conclusion that basically it is not the mathematical technique that is at fault, but rather the theory itself is not logically consistent.

Within the framework of the present theory, it is still possible to calculate many results, namely, those which do not depend critically on the assumed size of the particle. A few years ago, important new successes in this direction were obtained by Tomanaga[16], Schwinger[17], and Feyman[18], with the prediction of certain very fine details of the spectrum of hydrogen gas, as well as with experiments that measured the magnetic moment of the electron. Impressive as these results are, considered as examples of extraordinarily complex calculations that led to correct experimental predictions, they throw little light, however, on the problem of the infinities that is one of the most important manifestations of the current crisis in physics. For a closer examination of these calculations shows that they do not depend significantly on what happens at distances that are much shorter than the Compton wave-length of the electron (about 3×10^{-11} cm.), while other considerations which we shall discuss presently suggest that the failure of current theories should first become important around 10^{-13} cm. The agreement of these calculations with experiment then constitutes an excellent verification of the current quantum theory in the domain in which all other indications suggest that it ought to be valid. However, it has also become clear that because this kind of experiment is so insensitive to the details of what happens in the domain of very short distances, it does not provide a very promising tool for investigating this domain.

On the other hand, experiments with particles of very high energy (of the order of 100 million electron volts or more) have led to a bewildering array of new phenomena, for which there is no adequate treatment in the existing theory. For as we have already pointed out in previous chapters, one discovers that the so-called "elementary particles" of physics, such as protons and neutrons, can now transform into each other. Moreover, many new particles, the positron, the neutrino, about ten different kinds of "mesons" and several new

---

* Most of the difficulties originate in connection with making such an assumption consistent with the theory of relativity.

kinds of particles called hyperons have been discovered. No visible limit to this process of discovering new particles appears to be in sight as yet. And most of these new particles are unstable, having the ability to transform into each other, and to "decay" ultimately into neutrons, electrons, and protons. Besides, they can all be "created" in energetic collisions of other particles with nuclei. Moreover, a more accurate analysis of the data suggests that these new properties of matter become important only when particles approach within a distance of each other that is of the order of 10^{-13} cm. or less. Hence, in experiments carried out at the atomic level, practically no indication of these new properties is to be found.

It is evident, then, that the entire scheme by which the universe is regarded as made of certain kinds of "elementary particles" has demonstrated its inadequacy, and that some very different concept is needed here. Thus, when a similar instability and transformability of atoms in radioactive transformations was discovered half a century ago, it soon became evident that the chemical "elements" were not really elementary, being composed in fact, as was discovered later, of protons, electrons, and neutrons. Similarly, it seems reasonable to conclude that in the domain of very high energy experiments, we are disturbing the present-day "elementary" particles sufficiently so that their actual structure is beginning to manifest itself. According to the considerations that we have discussed previously, this structure should have a size of the order of 10^{-13} cm.

It is easy to see that there are strong reasons for supposing a connection between the problem of the structure of the "elementary" particles and that of the infinities predicted by current theories. For if particles have a structure, this already implies that they occupy some space. And if they occupy space, then they will not be mathematical points, so that there will be no occasion for these infinities to arise. Just what the internal structure of these particles is we do not know as yet, but to find out is now our problem. Evidently, it must be something new relative to what is known so far. In the next section we shall discuss the indications that now exist regarding the nature of this structure.

8. ADVANTAGES OF NEW INTERPRETATION OF QUANTUM THEORY IN THE GUIDANCE OF RESEARCH IN NEW DOMAINS

We shall now discuss some of the principal advantages of the suggested new interpretation of the quantum theory over the usual interpretation in the guidance of research aimed at resolving this crisis.

First of all, let us recall that one of the principal problems now faced in this domain is that of treating the structure of an "elementary" particle, and of discovering what kinds of motions are taking place within this structure—motions that would help explain, perhaps, the "creation" and "destruction" of various kinds of particles, and their transformation into each other. In the usual interpretation of the quantum theory, it is extraordinarily difficult to consider this problem. For the insistence that one is not to be allowed to conceive of what is happening at this level means that one is restricted to making blind mathematical manipulations with the hope that somehow one of these manipulations will lead to a new and correct theory.

Secondly, the usual interpretation of the quantum theory implies a certain general mathematical and physical scheme which does not seem to lend itself very well to the notion that matter has new kinds of properties connected with an inner structure of the "elementary" particles. This general scheme, which we have already mentioned in Chapter III, Section 9, is the one involving purely linear equations for a wave function in a configuration space, "observables" obtained from linear operators, a purely probabilistic interpretation of the wave functions, etc. If one adopts this scheme, then the only mathematical possibilities left open seem to be the modification of current versions of the quantum theory, by alterations of the equations in such a way as somehow to cut out the contributions from short distances that lead to the infinities. Throughout the past twenty years, a great deal of intensive research has been devoted to attempts to do this in various ways (by cut-offs, finite distance operators, S-matrices, etc.), but none of these efforts has as yet shown any promise of leading to a consistent theory. These attempts have in general been guided by the expectation, commonly held, as we have seen in Section 1, by modern theoretical physicists, that in future theories the behaviour of things will be even less precisely definable than is possible in current theories. Of course, it cannot be proved at present that these expectations are definitely wrong. But the failure of the large number of efforts that have already been made in this direction would suggest that it may well be fruitful to try other lines of approach, especially considering that, as we have seen in the previous chapter, the restriction to the currently accepted line of approach is in any case not justifiable by any experimental or theoretical evidence coming from physics itself.

On the other hand, when we attack these problems within the framework of the new interpretation of the quantum theory, a large number of interesting new possibilities are seen to open up. First of all, the work is considerably facilitated by the fact that we can imagine what is happening, so that we can be led to new ideas not only by looking directly for new equations, but also by a related procedure of thinking in terms of concepts and models that will help to suggest new equations, which would be very unlikely to be suggested by mathematical methods alone. More important, however, is the fact that in terms of the notion of a sub-quantum mechanical level, we are enabled to consider a whole range of qualitatively new kinds of theories, approaching the usual form of the quantum theory only as approximations that hold in limiting cases. Moreover, there are a number of reasons, suggesting that the new features of such theories are likely to be relevant in the treatment of processes involving very high energies and very short distances. Some of these reasons are:

(1) If there is a sub-quantum mechanical level, then, as we saw in Section 2, processes with very high energy and very high frequency may be faster than the processes taking place in the lower level. In such cases, the details of the lower level would become significant, and the current formulation of the quantum theory would break down.

For example, in our point of view, the "creation" of a particle, such as a meson, is conceived as a well-defined process taking place in the sub-quantum mechanical level. In this process, the field energy is concentrated in a certain region of space in discrete amounts, while the "destruction" of the particle is just the reverse process, in which the energy disperses and takes another form. In the quantum domain, the precise details of this process are not significant, and can therefore be ignored. This is in fact what is done in the current quantum theory which discusses the "creation" and "destruction" of particles as merely a kind of "popping" in and out of being in a way that is simply not supposed to be subject to further description.* With very fast high energy processes, however, the results may well depend on these details, and if this should be the case, the current quantum theory would not be adequate for the treatment of such processes.

(2) The treatment of the Dirac quantum in our theory leads, as we saw in the previous section, to the possibility of describing many different kinds of fields in terms of different modes of vibration of a single basic field. Moreover, in very energetic processes, the approximation that reduces to the usual form of the quantum theory would break down. This fact would have the further desirable consequences that the infinities characteristic of the current quantum theory could be eliminated. For in our theory it can be seen that in an exact

---

* This is done mathematically with the aid of the so-called creation and destruction operators, which describe the coming into existence of a given kind of particle in a certain state by the bare statement that the number of such particles has increased by unity, while its passing out of existence is described by saying that this number has decreased by unity.

treatment, no results can become infinite. Thus, we are led to interpret the infinities as a consequence of an unjustified extrapolation of all the features of the current quantum theory to very short distances and to very high energies.

It can be seen that the assumption that the usual form of the quantum theory will continue to hold in the high energy domain is equivalent to the assumption that there is no sub-quantum mechanical level, or at least, if there is such a level, we have not yet reached a point where its effects are significant. It is of course possible that this assumption may be true. But enough evidence has been shown favouring the plausibility of the opposite assumption that it no longer seems to be justifiable to restrict *all* research in theoretical physics to those lines that fit into a continuation of the usual interpretation of the quantum theory.

9. ALTERNATIVE INTERPRETATION OF THE QUANTUM THEORY AND THE PHILOSOPHY OF MECHANISM

The consideration of the alternative interpretation of the quantum mechanics discussed in this chapter serves to show that when one divests the theory of the irrelevant and unfounded hypotheses of the absolute and final validity of the indeterminacy principle, one is led to an important new line of development, which strikes at the roots of the entire mechanist philosophy. For we now see that there is a *whole level* in which chance fluctuations are an inseparable part of the mode of being of things, so that they must be interwoven into the fabric of the theory of this level in a fundamental way. Thus, we have been led to take an important step beyond the classical notion of chance as nothing more than the effects of contingencies that modify the boundary conditions or introduce randomly fluctuating external forces in a way that is not predictable within the context of interest, but which play no essential part in the formulation of the basic laws that apply within such a context.

If we stopped at this point, however, we should, as we have seen in the previous chapter, merely have switched from deterministic to indeterministic mechanism. To avoid indeterministic mechanism, we must suppose that, in their turn, the chance fluctuations come from something else. Since, as Heisenberg and Bohr have shown so well, there is no room in the quantum domain for anything to exist in which these fluctuations might originate, it is clear that to find their origin we must go to some new domain. As we have seen in this chapter, there is much evidence suggesting the plausibility of the notion that they originate in a sub-quantum mechanical level. Nevertheless, independently of the specific proposals that we have made here, the essential point with regard to the question of mechanism is that the fluctuations should come from qualitatively new kinds of factors existing in a new domain.

Within the new domain described above, we would naturally expect that new kinds of laws would operate, which may include new kinds of causal laws as well as new kinds of laws of chance. Of course, if one were now to make the assumption that these new laws would surely be nothing more than purely causal laws, one would then fall back into deterministic mechanism, while the similar assumption that they were surely nothing more than laws of probability would throw one back into indeterministic mechanism. On the other hand, we have in the proposals made in this chapter avoided both these dogmatic and arbitrary extremes, since we have considered, as the situation demanded, the possibility that there are new features to the causal laws (a "quantum force" not appearing at higher levels) as well as to the laws of chance (random fluctuations originating in the sub-quantum mechanical level).

Of course, as we have indicated in Section 5, we do not regard our earlier proposals as providing a completely satisfactory and definitive interpretation of the laws of the quantum domain. The basic reason is, in a sense, that the fundamental concepts considered in the theory (waves and particles in interaction) are still very probably too close to those applying in the classical domain to be appropriate to a completely new domain such as that treated in the quantum theory. Indeed, the whole general effort to understand the quantum theory in terms of models so close to those of the classical domain has often been criticized as mechanistic. This criticism would in fact be correct if one's intention were to stop at this point. On the other hand, if one simply regards these theories as something definite from which it may be helpful to start, then it seems evident that such a criticism does not apply.

It is important to add here that there are good reasons why the provisional consideration of mechanical explanations of the quantum theory may be a good starting-point from which qualitatively new developments are likely to arise.

First of all, one must recall that mechanical theories often imply qualitatively new properties. Thus, for example, when a large number of simple mechanical elements is put into interaction one obtains fundamentally new kinds of collective effects (e.g. the large-scale properties of an aggregate of atoms). Thus, we may expect that the consideration of old mechanical concepts in new contexts will perhaps already lead to some qualitatively new results.

Secondly, a careful consideration of the contradictions and the weak points of a given mechanical theory often suggests new concepts that resolve these contradictions or eliminate the weak points.

Thus, the careful analysis of the mechanical theory of the electrons made by Lorentz was of great help in suggesting the theory of relativity, which gave a solution to many of the difficulties raised by the Lorentz theory. In Section 6, we have described efforts to deal with some of the weak points of our original interpretation, which likewise suggest considerable changes relative to the original conception.

Thirdly, to insist that once we pass the classical domain, mechanical concepts will surely have absolutely no relevance whatever would be just as dogmatic as to insist that these concepts must be adequate for every domain that will ever be investigated. A better procedure is to try every kind of concept that we can think of, and to see which kind is best in each particular domain. The work described in this chapter then shows that mechanical concepts can go further in the quantum domain than had hitherto been thought possible.

Finally, it is important to stress the value of starting with some concrete theory and working forwards from there. Without such a concrete starting-point, criticism of the current theories is rather likely to become sterile in the long run. For it is extremely difficult purely from general considerations to be led to qualitatively new ideas. Thus, in practice, such criticism, accompanied by no concrete new suggestions, is likely to leave one with no real alternative but to continue to work along the usual lines, in the hope that new experimental developments or lucky and brilliant new theoretical insights will eventually lead to a new theory. On the other hand, to leave future progress in this line solely to experiment or to the hope of chance new insights means that one is renouncing one of the important functions of criticism, namely, to help suggest definite alternative lines of research that are likely to lead to a correct direction. And as we have pointed out here, there is good reason to suppose that the specific proposals indicated in this chapter may be helpful in achieving this purpose.

BIBLIOGRAPHY

List of References for Chapter IV

(1) R. Furth, *Zeits. f. Phys.*, **81,** 143 (1933).
(2) L. de Broglie, *Compt. Rend.*, **183,** 447 (1926); **185,** 380 (1927).
(3) E. Madelung, *Zeits. f. Phys.*, **40,** 332 (1926).
(4) *Reports on the Solvay Congress*, Gauthier-Villars, Paris (1928), p. 280.
(5) L. de Broglie, *The Revolution in Physics*, Routledge & Kegan Paul, London (1954).

Bibliography

(6) *Uspekhi. fizich. Nauk*, **45** (Oct. 1951); French translation in *Questions Scientifiques*, Vol. 1 (Editions de la Nouvelle Critique, Paris (1952)). See also D. J. Blokhinzhev, *Grundlagen der Quantenmechanik*, Deutscher Verlag der Wissenschaften, Berlin (1953).

(7) D. Bohm, *Phys. Rev.*, **85**, 166 (1952).

(8) D. Bohm, *Phys. Rev.*, **85**, 180 (1952).

(9) Vigier's suggestions are discussed in L. de Broglie, *La Physique Quantique Restera-t-elle Indeterministe*, Gauthier-Villars, Paris (1953).

(10) D. Bohm, *Prog. Theor. Physics*, **9**, 273 (1953).

(11) D. Bohm and J. P. Vigier, *Phys. Rev.*, **96**, 208 (1954).

(12) D. Bohm, R. Schiller, and J. Tiomno, *Supplemento al Nuovo Cimento*, I, Serie X, 48 (1953).

(13) T. Takabayasi, *Prog. Theor. Physics*, **8**, 143 (1952); **9**, 187 (1953).

(14) I. Fenyes, *Zeits. f. Physik*, **132**, 81 (1952).

(15) W. Weizel, *Zeits f. Physik*, **134**, 264 (1953); **135**, 270 (1953).

(16) Fukada, Miyamoto and Tomanaga, *Prog. Theor. Physics*, **4**, 47 and 121 (1949).

(17) J. Schwinger, *Phys. Rev.*, **74**, 749 and 769 (1949); **80**, 440 (1950).

(18) R. P. Feyman, *Phys. Rev.*, **75**, 486 and 1736 (1949).

CHAPTER FIVE

More General Concept of Natural Law

1. INTRODUCTION

WE have seen throughout this book that for several centuries there has existed a very strong tendency for one form or another of the philosophy of mechanism to be generally adopted among physicists. In Chapters II and III we have described the essentials of this philosophy in some detail, and have given a general outline of how this philosophy has developed in response to the new problems with which physics was faced during the nineteenth and twentieth centuries. In the present chapter we shall criticize this philosophy, demonstrating the weaknesses in its basic assumptions, and then we shall go on to propose a different and broader point of view which we believe to correspond more nearly than does mechanism to the implications of scientific research in a wide range of fields. In addition to presenting this broader point of view in some detail, we shall also show how it permits a more satisfactory resolution of several important problems, scientific as well as philosophical, than is possible within the framework of a mechanistic philosophy.

2. SUMMARY OF THE ESSENTIAL CHARACTERISTICS OF A MECHANISTIC PHILOSOPHY

The essential characteristics of a mechanistic philosophy in the most general form that it has developed thus far in physics are the following:

The enormous diversity of things found in the world, both in common experience and in scientific research, can *all* be reduced completely and perfectly and unconditionally (i.e. without approximation and in every possible domain) to nothing more than the effects of some definite and limited general framework of laws. While it is admitted that the details of these laws may be subjected to changes in accordance with new experimental results that may be obtained in the future, its basic general features are regarded as absolute and final. This means that the fundamental entities that are supposed to exist, the kinds of qualities that define the modes of being of these entities, and the general kinds of relationships in terms of which the basic laws are to be expressed, are supposed to fit into some fixed and limited physical and mathematical scheme, which could in principle be subjected to a complete and exhaustive formulation, if indeed it is not supposed that this has already been done. At bottom, the only changes that are regarded as possible within this scheme are quantitative changes in the parameters or functions defining the state of the system (as precisely as the nature of the system permits this state to be defined),* while fundamental qualitative changes in the modes of being of the basic entities and in the forms in which the basic laws are to be expressed are not regarded as possible. Thus, the essence of the mechanistic position lies in its assumption of fixed basic qualities, which means that the laws themselves will finally reduce to purely quantitative relationships.

As we have seen in previous chapters, the philosophy of mechanism has undergone an extensive evolution in its specific form, all the while retaining the essential characteristics described above, in forms that tend, however, to become more and more complex and subtle with the further development of science.

3. CRITICISM OF THE PHILOSOPHY OF MECHANISM

We shall now review some of the most important criticisms that can be made against the philosophy of mechanism.

First of all, the historical development of physics has not confirmed the basic assumptions of this philosophy, but rather, has continually contradicted them. Thus, since the time of Newton, there have been introduced, not only the whole series of specific changes in the conceptual structure of physics† that was discussed in Chapter II, but also the revolutionary changes in the whole general framework, brought about by the theory of relativity and the quantum theory.‡ Moreover, physics is now faced with a crisis in which it is generally admitted that further changes will have to take place, which will probably be as revolutionary compared to relativity and the quantum theory as these theories are compared to classical physics.

Secondly, the mechanistic assumption of the absolute and final character of any feature of our theories is never necessary. For the possibility is always open that such a feature has only a relative and limited validity, and that the limits of its validity may be discovered

* For example, in the usual interpretation of the quantum theory, the state of a system is subject, in general, only to a statistical determination.

† The field concept, the concept of quantitative changes that lead to qualitative changes, the concepts of chance and statistical law.

‡ These changes, especially those resulting from the quantum theory, have been discussed mainly in Chapters III and IV.

in the future. Thus, Newton's laws of motion, regarded as absolute and final for over two hundred years, were eventually found to have a limited domain of validity, these limits having finally been expressed with the aid of the quantum theory and the theory of relativity. Indeed, as we saw in more detail in Chapter II, Sections 13 and 15, the mechanistic thesis that certain features of our theories are absolute and final is an assumption that is not subject to any conceivable kind of experimental proof, so that it is, at best, purely philosophical in character.

Thirdly, the assumption of the absolute and final character of any feature of our theories contradicts the basic spirit of the scientific method itself, which requires that *every* feature be subjected to continual probing and testing, which may show up contradictions at any point where we come into a new domain or to a more accurate study of previously known domains than has hitherto been carried out. Indeed, the normal pattern that has developed without exception in every field of science studied thus far has been just the appearance of an endless series of such contradictions, each of which has led to a new theory permitting an improved and deeper understanding of the material under investigation. Thus, the full and consistent application of the scientific method makes sense only in a context in which we refrain from assuming the absolute and final character of any feature of any theory and in which we therefore do not accept a mechanistic philosophy.

Of course, the above arguments do not prove that a mechanistic philosophy is definitely wrong. For it is always conceivable that the trouble thus far has been that we have just not found the true absolute and final theory, and that this theory may be somewhere beyond the horizon of current scientific research. On the other hand, the historically demonstrated inadequacy of this philosophy up to the present, the fact that its basic assumptions cannot possibly be proved, and the fact that they are in disagreement with the whole spirit of the scientific method, would suggest to us that it may well be worth our while to consider points of view that go outside the limits of a mechanistic philosophy. It is with the development of a point of view having such an aim that we shall be concerned throughout the rest of this chapter.

4. A POINT OF VIEW THAT GOES BEYOND MECHANISM

The nucleus of a point of view that goes beyond mechanism and that is also in better accord than is mechanism with general scientific experience and with the needs of scientific research has already been presented in Chapter I, Section 10 and in Chapter II, Section 15, in connection with the extremely rich and diversified structure that has thus far actually been found in the laws of nature. The most essential feature characterizing this general structure is this: Any given set of qualities and properties of matter and categories of laws that are expressed in terms of these qualities and properties is in general applicable only within limited contexts, over limited ranges of conditions and to limited degrees of approximation, these limits being subject to better and better determination with the aid of further scientific research. Indeed, both the very character of the empirical data and the results of a more detailed logical analysis show that beyond the above limitations on the validity of any given theory, the possibility is always open that there may exist an unlimited variety of additional properties, qualities, entities, systems, levels, etc., to which apply correspondingly new kinds of laws of nature. Or, lumping all of the above diverse possibilities into the single category of "things", we see that a systematic and consistent analysis of what we can actually conclude from experimental and observational data leads us to the notion that nature may have in it an infinity of different kinds of things.

It is clear that this point of view carries us completely outside the scope of what can be considered a mechanistic philosophy. For, as we recall, the mechanistic point of view involves the assumption that the possible variety in the basic properties and qualities existing in nature is limited, so that one is permitted at most to consider quantitative infinities, which come from making some finite number of kinds of things bigger and bigger or more and more numerous. Moreover, it is also clear that the notion to which we have been led is quite distinct from that of a series of successive approximations that converge to some fixed and limited set of final laws. For there is evidently no reason why new qualities and properties and the corresponding new laws should *always* lead just to smaller and smaller corrections that converge in this simple and uniform way towards definite results. This may well be what happens in certain contexts and within a definite range of conditions. Nevertheless, there is no conceivable empirical justification for excluding the possibility that in different contexts or under changed conditions these new qualities, properties, and laws will lead to effects that are not small in relation to those following from previously known properties, qualities, and laws. Thus, for example, while the laws of relativity and quantum theory do in fact lead under special conditions to small corrections to those of Newtonian mechanics, they lead more generally, as is well known, to qualitatively new results of enormous significance, results that are not contained in Newtonian mechanics at all.* The same possibility evidently

* E.g. the "rest energy" of matter, the stability of atoms, etc.

necessarily exists with regard to any other new laws that may eventually be discovered. Therefore, the assumption that the laws of nature constitute an infinite series of smaller and smaller steps that approach what is in essence a mechanistic limit is just as arbitrary and unprovable as is the assumption of a finite set of laws permitting an exhaustive treatment of the whole of nature.

We see then, that, as far as the empirical data of science themselves are concerned, they cannot justify any *a priori* restrictions at all, either on the character or on the relative importance in different conditions and contexts of the inexhaustibly rich and diversified qualities and properties that may exist in nature. Such qualities and properties—which can always, as far as we are able to tell, lie hidden behind the errors and inadequacies of any given set of theories—may be disclosed later in an investigation carried out under new conditions, in new contexts, or to new degrees of approximation.

Thus far, we have been led by our analysis of the character of empirical data and of scientific theories only to a consideration of the *possibility* that nature may have in it an infinity of potentially or actually significant qualities (i.e. qualities which are of major importance or which can become of major importance under suitable conditions and in suitable contexts). It is now clear, however, that there are really only two possibilities with regard to this problem. Either the qualities having this kind of significance are limited in number, or else they are not. To suppose the former is essentially to fall back into one form or another of the mechanistic philosophy, to which, as we have seen, so many objections can be raised. If we wish to go outside the mechanistic philosophy, we therefore really have no choice but to consider the consequences of the assumption that the number of such significant qualities is not limited.

We have thus been led to see what is the first crucial step towards a point of view that goes beyond the mechanistic philosophy. On the other hand, at this stage of the analysis, this point of view presents itself as one of two possible alternatives: i.e. either mechanism or an infinity of potentially or actually significant qualities. Evidently we must choose one or the other. But on what basis can we make such a choice? In order to answer this question we point out that the notion of the qualitative infinity of nature becomes more than merely an alternative to the philosophy of mechanism, if we take into account the rôle of conditions, context, and degree of approximation in limiting the domain of applicability of any given theory. For, with this addition, it constitutes a broader point of view, in the sense that it contains within it all of those consequences of mechanism which represent a genuine contribution to the progress of scientific research, while it does not contain those which make no such contribution and which impede scientific research. To see this, we first note that, with regard to any given domain of phenomena, the specific form of the assumption of the qualitative infinity of nature that has been suggested above does not contradict the notion that these phenomena can be treated in terms of some finite set of qualities and laws, and indeed, in terms of a number much smaller than the number of items of empirical data that may be available. It is evident that if this were not possible, then one of the most important achievements of scientific theories would be lost, for they would no longer permit the explanation* and prediction of a large number of at first sight independent phenomena on the basis of relatively few general qualities, properties, laws, principles, etc. The recognition of this possibility and its practical exploitation in a wide range of fields was indeed the basic contribution that the mechanistic philosophy brought to science in the early phases of its development.† As we have seen, however, as long as we qualify our theories by specifying the context, conditions, and degrees of approximation to which they are valid, or at least by admitting that these limitations on their validity must eventually be discovered, then the notion of the qualitative infinity of nature leads one to treat any given domain of phenomena in exactly the same way as is done if one adopts a mechanistic point of view. It is only with regard to predictions in new domains, in new contexts, and to new degrees of approximation that the qualitative infinity of nature dictates an additional measure of caution, since it implies that eventually (but exactly where must be determined only empirically) any limited number of qualities, properties, and laws will prove to be inadequate. But, as we have seen, the very form of the data themselves, as well as a logical analysis of their meaning, dictates exactly the same measure of caution. We see, then, that none of the really well-founded conclusions that can be obtained with the aid of the assumption of a finite number of qualities in nature can possibly be lost if we assume instead that the number of such qualities is infinite, and at the same time recognize the rôle of contexts, conditions, and degrees of approximation.‡ All that we can lose is the illusion that

---

* Let us recall that, as pointed out in Chapter I, Section 3, explanation constitutes showing that certain things follow necessarily from others.

† See, for example, Chapter II, Section 3.

‡ It is by recognizing that a finite and generally limited number of qualities, properties, and laws may be adequate in given contexts, conditions, and degrees of approximation that we avoid the procedure of simply falling back into an arbitrary multiplication of qualities that was characteristic of the pre-mechanistic point of view, especially in the scholastic form of the Aristotelian philosophy that was prevalent in the Middle Ages.

we have good grounds for supposing that in principle we can, or eventually will be able, to predict everything that exists in the universe in every context and under all possible conditions.

Not only can nothing of real value for scientific work be lost if we adopt the notion of the qualitative infinity of nature in the specific form that has been described here, but on the contrary, much can be gained by doing this. For, first of all, we can thereby free scientific research from irrelevant restrictions which tend to result from (and which have in fact so often actually resulted from) the supposition that a particular set of general properties, qualities, and laws must be the correct ones to use in all possible contexts and conditions and to all possible degrees of approximation. Secondly, we are led to a concept of the nature of things which is in complete accord with the most basic and essential characteristic of the scientific method; i.e. the requirement of continual probing, criticizing and testing of every feature of every theory, no matter how fundamental that theory may seem to be. For this view explains the necessity for doing scientific research in just this way and in no other way, since, if there is no end to the qualities in nature, there can be no end to our need to probe and test all features of all of its laws. Finally, as we shall show throughout the rest of this chapter, the assumption of the qualitative infinity of nature leads to a much more satisfactory solution of a number of important problems, both scientific and philosophical, than is possible within the framework of a mechanist philosophy; and this in turn gives further evidence that it is a better point of view for the guidance of scientific research.

In conclusion, then, the notion of the qualitative infinity of nature permits us to retain all the positive achievements that were made possible by the development of mechanism. In addition, it enables us to go beyond mechanism by showing the limitations of the latter philosophy and by pointing towards new directions in which our concepts and theories may undergo further development. Naturally, we do not wish to propose here that the qualitative infinity of nature is a final doctrine, beyond which no further steps can ever be made. Indeed, as science progresses, it seems very likely that the qualitative infinity of nature will eventually be found to fit into some still more general point of view, which in turn retains its positive achievements, and which goes beyond them, much as the motion of the qualitative infinity of nature goes beyond mechanism. But, in this chapter, our purpose is merely to call attention to the many factors that suggest the need for this important step carrying us outside the limits of a mechanistic philosophy, and to show the numerous advantages that come from taking this step.

5. MORE DETAILED EXPOSITION OF THE MEANING OF QUALITATIVE INFINITY OF NATURE

In this section we shall bring out in more detail what is the general view of the world implied by the notion of the qualitative infinity of nature, and we shall show how this view agrees with the actual results of research that have been obtained thus far in the field of physics.

In order to make possible a discussion in relatively concrete terms we shall begin by considering a specific example: viz. the atomic theory of matter. Now, as is well known, the earliest forms of the atomic theory were based on the assumption that the fundamental qualities and properties defining the modes of being of the atoms were limited in number. On the other hand, as we have pointed out many times, deeper studies of the atom have disclosed more and more details of a moving substructure, which has within it a richness of properties and qualities that has never yet shown the slightest sign of being exhausted or of approaching exhaustion. Thus, there was found in the atoms a structure of electrons moving around a central nucleus consisting of neutrons and protons which themselves took part in further characteristic kinds of motions of their own. Within all of these motions appeared quantum-mechanical fluctuations of various kinds. Then came the discovery of a structure for the electrons and protons involving in some as yet poorly understood way the motions of unstable particles such as mesons and hyperons. Still later came the realization that because these latter particles can be "created", "destroyed", and transformed into each other, they too are very likely to have a further structure that is related to the motions of some still deeper-lying kinds of entities the nature of which is not yet known.

An essential characteristic of the rich and highly interconnected substructure of moving matter described above is that not only do the quantitative properties in it continually change but that the basic qualities that define its mode of being can also undergo fundamental transformations when conditions alter sufficiently. Thus, in electrical discharges, atoms can be excited and ionized, in which case they obtain many new physical and chemical properties. Under bombardment with very high-energy particles, the nuclei of the various chemical elements can be excited and transformed into new kinds of nuclei, with even more radical changes in their physical and chemical properties. Moreover, in nuclear processes, neutrons can be transformed into protons, either by the emission of neutrinos or of mesons; and of course, as we have seen, mesons are unstable, so that their very mode of existence implies the necessity for their

137

transforming into basically different kinds of particles. Thus, further research into the structure of matter has not only shown what is, as far as we have been able to tell, an unlimited variety of qualities, processes, and relationships, but it has also demonstrated that all of these things are subject to fundamental transformations that depend on conditions.

Thus far, we have tended to emphasize the inexhaustible *depth* in the properties and qualities of matter. In other words, we have considered how experiments have shown the existence of level within level of smaller and smaller kinds of entities, each of which helps to constitute the substructure of the entities above it in size, and each of which helps to explain, at least approximately, by means of its motions how and why the qualities of the entities above it are what they are under certain conditions, as well as how and why they can change in fundamental ways when conditions change. But now we must take into account the fact that the basic qualities and properties of each kind of entity depend not only on their substructures but also on what is happening in their general background. In physics, research thus far has not tended to stress this feature of the laws of nature as much as it has emphasized the substructure. Nevertheless, the various fields (e.g. electromagnetic, gravitational, mesonic, etc.) that have been introduced into the conceptual structure of physics represent to some extent an explicit recognition of the importance of the background. For, as we have seen, these fields (whose mode of existence requires that they be defined over broad regions of space) enter into the definition of the basic characteristics of all the fundamental particles of current physics. Moreover, when such fields are highly excited, they too can give rise to qualitative transformations in the particles, while, vice versa, the particles have an important influence on the character of the fields. Indeed, the discussion of the quantum theory given in Chapters III and IV shows that fields and particles are closely linked in an even deeper way, in the sense that both are probably opposite sides of some still more general type of entity, the detailed character of which remains to be discovered.* Thus, the next step in physics may well show the inadequacy of the simple procedure of just going through level after level of smaller and smaller particles, connected perhaps by fields which interact with these particles. Instead, we may find that the background enters in a very fundamental way even into the definition of the conditions for the existence of the new kinds of

---

* This is suggested by the wave-particle duality in the general properties of matter, which implies, as we have seen, that we may have to deal with some new kind of thing that can, under suitable circumstances, act either like a localized particle or like an extended field.

basic entities to which we will eventually come, whatever they may turn out to be. Thus, we may be led to a theory in which appears a much closer integration of substructure and background into a well-knit whole than is characteristic of current theories.

We see from the above discussion that the qualitative infinity of nature is not equivalent to the idea expressed by the well-known rhyme:

> "Great fleas have little fleas
> Upon their backs to bite 'em;
> Little fleas have lesser fleas,
> And so *ad infinitum*."

For, firstly, we are not supposing that the same pattern of things is necessarily repeated at all levels; and secondly, we are not even supposing that the general pattern of levels that has been so widely found in nature thus far must necessarily continue without limit. While we cannot decide this question from what is known at present, we have already suggested reasons why we may perhaps now be approaching a point at which the notion of levels will, at the very least, have to be enriched a great deal by the explicit inclusion of the effects of a background that is essential for the very existence of the entities in terms of which our theories are to be formulated. Moreover, it is evidently quite possible that as we penetrate further still, we will find that the character of the organization of things into levels will change so fundamentally that even the pattern of levels itself will eventually fade out and be replaced by something quite different. Hence, while the qualitative infinity of nature is consistent with an infinity of levels, it does not necessarily imply such an infinity. And, more generally, this notion does not require *a priori* the continuation of any special feature of the general pattern of things that has been found thus far, nor does it exclude *a priori* the possibility that any such feature may continue to be encountered, perhaps in new contexts and in new forms, no matter how far we may go. Such questions are left to be settled entirely by the results of future scientific research.

There is, however, one general statement that can be made at this point about the inexhaustible diversity of things that may exist in the universe; namely, that they must have some degree of autonomy and stability in their modes of being. Now, thus far, we have always found that such autonomy exists.* Indeed, if it did not exist, then we

---

* This autonomy may have many origins; e.g. the falling of the propagation of influences of one thing in another with an increase of separation between them, the decay of such influence with the passage of time, electrical screening, the existence of thresholds, such that influences

would not be able to apply the concept of a "thing" and there would then be no way even to formulate any laws of nature. For how can there be an object, entity, process, quality, property, system, level, or whatever other thing one cares to mention, unless such a thing has some degree of stability and autonomy in its mode of existence, which enables it to preserve its own identity for some time, and which enables it to be defined at least well enough to permit it to be distinguished from other things? If such relatively and approximately autonomous things did not exist, then laws would lose their essential significance (e.g. they could not in principle be tested by altering conditions with the aid of experiments in the manner described in Chapter I, Section 3, because the basic things entering into the laws would change all their characteristic modes of being with the slightest change of conditions).

In conclusion, then, actual scientific research has thus far shown the need to analyse nature in terms of a series of concepts that involve the recognition of the existence of more and more kinds of things; and the development of such new concepts has never yet shown any signs of coming to an end. Up to the present, the various kinds of things existing in nature have, at least as far as investigations in the field of physics are concerned, been found to be organized into levels. Each level enters into the substructure of the higher levels, while, vice versa, its characteristics depend on general conditions in a background determined in part in other levels both higher and lower, and in part in the same level. It is quite possible, of course, that further studies will disclose a still more general pattern of organization of things. In any case, it is clear that the results of scientific research to date strongly support the notion that nature is inexhaustible in the qualities and properties that it can have or develop. If the laws of nature are to be expressible in any kind of terms at all, however, it is necessary that the things into which it can be analysed shall have at least some degree of approximate and relative autonomy in their modes of being, which is maintained over some range of variation of the conditions in which they exist.

6. CHANCE AND NECESSARY CAUSAL INTERCONNECTIONS

With the aid of the general world view described in Section 5, we shall now proceed to show that the hypothesis of the qualitative

---

which are too weak to surpass these thresholds produce no significant effects; the fact that individual constituents of an object (such as atoms) are too small to have an appreciable effect on the object as a whole, while collectively there is a considerable independence of motions of the constituents leading to the cancellation of chance fluctuations. Many other such sources of autonomy exist, and doubtless more will be discovered in the future.

infinity of nature provides a framework within which can fit quite naturally the concept, suggested in Chapter I, of chance and necessary causal interconnections as two sides of every real natural process.

First of all, we point out that if there are an unlimited number of kinds of things in nature, no system of purely determinate law can ever attain a perfect validity. For every such system works only with a finite number of kinds of things, and thus necessarily leaves out of account an infinity of factors, both in the substructure of the basic entities entering into the system of law in question and in the general environment in which these entities exist. And since these factors possess some degree of autonomy, one may conclude from the principle discussed in Chapter I, Section 8, that the things that are left out of any such system of theory are in general undergoing some kind of a random fluctuation. Hence, the determinations of any purely causal theory are always subject to random disturbances, arising from chance fluctuations in entities, existing outside the context treated by the theory in question. It thus becomes clear why chance is an essential aspect of any real process and why any particular set of causal laws will provide only a partial and one-sided treatment of this process, which has to be corrected by taking chance into account.*

Of course, it should not be supposed that every inadequacy or breakdown of causal laws must necessarily be due to the effects of chance fluctuations. Indeed, as happened in connection with the experiments leading to the theory of relativity (Michelson-Morley experiment, etc.), the failure of a given set of causal laws may represent just a simple and reproducible deviation between the predictions of these laws and the experimental results. A deviation of this kind implies only that the causal laws in question must be replaced by newer, more extensive, and more accurate causal laws (as indeed happened with Newtonian mechanics, which was replaced by the more general and more nearly correct relativistic mechanics). Quite often, however, experiments have disclosed not just simple and reproducible deviations from the predictions of a certain set of causal laws, but rather a breakdown of the entire scheme by which a specified set of properties are found to be related in a unique and necessary way in terms of a set of causal laws of a given general kind. Such a breakdown manifests itself in the appearance of chance fluctuations, not coming from anything in the context under investigation, but coming rather from qualitatively different kinds of fac-

---

*More generally, causal laws must be corrected by taking into account contingencies (see Chap. I, Sec. 8); because of the complex, multifold and interconnected character of these contingencies, however, their average effects can, in a wide range of conditions, be treated in terms of chance fluctuations and the theory of probability.

tors existing in contexts that are new relative to the one under consideration.* In such a case, the original causal law is seen to be valid only to the extent that the chance fluctuations in question cancel out, while in any given application the law will have a certain characteristic minimum range of error. This range of error is an objective property of the law in question, a property that is determined by the magnitudes of the chance fluctuations arising outside the context under investigation.

Vice versa, however, the characteristic limitation on the domain of validity of any given causal law which results from the neglect of the effects of chance fluctuations is balanced by a corresponding limitation on the domain of validity of any given law of chance, which results from the neglect of systematic causal interconnections between different contexts. In many cases (e.g. throws of a die) these interconnections are so unimportant that they have never yet been significant in any real applications. Nevertheless, this need not always be so. Consider, for example, the case of insurance statistics.. Here, one is able to make approximate predictions concerning the mean lifetime of an individual in a given group (e.g. one of definite age, height, weight, etc.) without the need to go into a detailed investigation of the multitudes of complex factors that contribute to the life or death of each individual in this group. This is possible only because the factors responsible for the death of any individual are extremely manifold and diverse, and because they tend to work more or less independently in such a way as to lead to regular statistical laws.† But the assumption underlying the use of these statistical laws are not always true. Thus, in the case of an epidemic or a war, the systematic interconnection between the cause of death of different individuals grows so strong that statistical predictions of any kind become practically impossible. To apply the laws of chance uncritically, by ignoring the possibility of corrections due to causal interconnections that may be unimportant in some conditions but crucially important in others, is therefore just as capable of leading to erroneous results as is the uncritical application of causal laws, in which one ignores the corrections that may be due to the effects of chance fluctuations.

A point of view that avoids the errors that generally result from assuming either causal laws or laws of chance to be basic and final kinds of laws is that suggested in Chapter II, Section 15. In this

---

* This is, for example, what happens to classical physics. For a particle such as an electron follows the classical orbit only approximately, and in a more accurate treatment is found to undergo random fluctuations in its motions, arising outside the context of the classical level (see Chapters III and IV).

† See, for example, Chapter I, Section 8.

point of view we regard both classes of laws as approximations, in the sense that just as a causal law can arise as a statistical approximation to the average behaviour of a large aggregate of elements undergoing random fluctuations, a law of chance can arise as a statistical approximation to the effects of a large number of causal factors undergoing essentially independent motions.* Actually, however, neither causal laws nor laws of chance can ever be perfectly correct, because each inevitably leaves out some aspect of what is happening in broader contexts. Under certain conditions, one of these kinds of laws or the other may be a better representation of the effects of the factors that are dominant and may therefore be the better approximation for these particular conditions. Nevertheless, with sufficient changes of conditions, either type of law may eventually cease to represent even what is essential in a given context and may have to be replaced by the other. Thus, we are led to regard these two kinds of laws as effectively furnishing different views of any given natural process, such that at times we may need one view or the other to catch what is essential, while at still other times, we may have to combine both views in an appropriate way. But we do not assume, as is generally done in a mechanistic philosophy,† that the whole of nature can eventually be treated completely perfectly and unconditionally in terms of just one of these sides, so that the other will be seen to be inessential, a mere shadow, that makes no fundamental contribution to our representation of nature as a whole. Thus, the notion of the qualitative infinity of nature leads us to the necessity of considering the laws of nature both from the side of causality and from that of chance, as well as more generally from new directions that may go beyond these two limits.

7. RECIPROCAL RELATIONSHIPS AND THE APPROXIMATE AND RELATIVE CHARACTER OF THE AUTONOMY OF THE MODES OF BEING OF THINGS

The qualitative infinity of nature has an important bearing on the problem of the reciprocal relationships between things, and on the question of the extent to which the modes of being of different things have an approximate autonomy.

First of all, we note that the universal interconnection of things has long been so evident from empirical evidence that one can no longer even question it. However, in a mechanistic point of view, it is assumed that this interconnection can ultimately be reduced to

---

* As we pointed out in Chapter II, Section 14, both these possibilities can be demonstrated mathematically, as well as with the aid of more qualitative types of arguments.
† See Chapter II, Sections 13 and 15.

nothing more than *interaction* between the fundamental entities which compose the system. By this we mean that in the mutual action of these entities on each other, there can only be quantitative changes in their properties, while fundamental qualitative changes in their modes of being cannot take place, provided that these entities are really the basic ones out of which the system is composed. Thus, in Newton's laws of motion there is equality of action and reaction of the elementary particles on each other, but this action and reaction is not supposed to affect the properties of the particles in a fundamental way.

On the other hand, in terms of the notion of the qualitative infinity of nature, one is led, as we have seen in previous sections, to the conclusion that every entity, however fundamental it may seem, is dependent for its existence on the maintenance of appropriate conditions in its infinite background and substructure. The conditions in the background and substructure, however, must themselves evidently be affected by their mutual interconnections with the entities under consideration. Indeed, as we have shown in many examples, this interconnection can, under appropriate conditions, grow so strong that it brings about qualitative changes in the modes of being of every kind of entity known thus far.* This type of interconnection we shall denote by the name of *reciprocal relationship*, to distinguish it from mere interaction.

The question now follows quite naturally, "If everything is in this very fundamental kind of reciprocal relationship with everything else, a relationship in which even the basic qualities and modes of being can be transformed, then how can we disentangle these relationships in such a way as to obtain an intelligible treatment of the laws governing the universe, or any part of it?" The answer is that all effects of reciprocal connections are not in general of equal importance. Of course we have the well-known fact already pointed out in Chapter I, Section 4, that within suitable contexts many of the reciprocal connections produce no significant effects, so that they can be ignored. On the other hand, if we consider a significant reciprocal connection between two things, then we must in general take both directions of this connection into account. If both directions are of comparable importance, then we will still find it very difficult to disentangle the real relationships between things, because one thing affects the basic qualities and laws determining the mode of being of the other; and this effect is returned in a complex process.

Experience in a wide range of fields of science shows, however, that both directions of a reciprocal connection do not always have to have comparable significance. When they do not have equal signifi-

* See, for example, Section 5 and Chapter II, Section 13.

144

cance, the problem is evidently simplified because the thing which has the major effect on the other is the dominant and controlling factor in the relationship. In this case we can study the laws and modes of being of the factors of major importance to a good degree of approximation, independently of the effects which may originate in the minor factor. A fundamental problem in scientific research is then to find what are the things* that in a given context, and in a given set of conditions, are able to influence other things without themselves being significantly changed in their basic qualities, properties, and laws. These are, then, the things that are, within the domain under consideration, autonomous in their essential characteristics to an adequate degree of approximation. When we have found such things, then we can make use of them for the prediction and control of the other things whose modes of being and basic characteristics are dependent on them. For example, in the case of the relationship between the large-scale level and the atomic level, we find, as pointed out in Chapter II, Section 13, that under conditions that are usually met and in most of the contexts that have thus far been treated in research in physics, the effect of the atomic motions on the laws of the large-scale level is much more important than the effects of the large-scale level on the laws of the atomic motions. Thus, it becomes possible by studying the laws of the atomic motions to make many kinds of approximate predictions concerning the laws and properties of things at the large-scale level and in this way to improve our understanding and control of the large-scale level.

On the other hand, as we saw also in Chapter II, Section 13, our prediction of the properties of the large-scale level through those of the atomic level can never be perfect, if only because there is a small but nevertheless real reciprocal influence of the large-scale level on the laws of the atomic level. This is due to the electronic and nucleonic substructure of the atom, which can be significantly affected by suitable conditions at the large-scale level (e.g. very high temperatures). Moreover, as we saw, the same possibility can arise with regard to the substructure of every entity that is known in physics (e.g. electrons, protons, mesons, etc.), provided that the conditions at the large-scale level are changed appropriately. As a result, we are led to the conclusion that in its reciprocal connections with the things existing in any *given* lower level, the entities at the macroscopic level must have at least some relative autonomy in their modes of being, in the sense that these modes cannot be predicted

---

* Let us recall that we are here using the word "thing" in a very general sense, so that it represents *anything* (e.g. objects, entities, qualities, properties, systems, levels, etc.).

perfectly from the specific lower level (or levels) in question. Even though the effects of this autonomy may be negligible in a wide range of conditions and contexts, it may nevertheless become very important in other conditions and contexts.

We see, then, that the existence of reciprocal relationships of things implies that each "thing" existing in nature makes some contribution to what the universe as a whole is, a contribution that cannot be reduced completely, perfectly, and unconditionally, to the effects of any specific set or sets of other things with which it is in reciprocal interconnection. And, vice versa, this also means evidently that no given thing can have a complete autonomy in its mode of being, since its basic characteristics must depend on its relationships with other things. The notion of a thing is thus seen to be an abstraction, in which it is *conceptually* separated from its infinite background and substructure. Actually, however, a thing does not and could not exist apart from the context from which it has thus been conceptually abstracted. And therefore the world is not made by putting together the various "things" in it, but, rather, these things are only approximately what we find on analysis in certain contexts and under suitable conditions.

To sum up, then, the notion of the infinity of nature leads us to regard each thing that is found in nature as some kind of abstraction and approximation. It is clear that we *must* utilize such abstractions and approximations if only because we cannot hope to deal directly with the qualitative and quantitative infinity of the universe. The task of science is, then, to find the right kind of things that should be abstracted from the world for the correct treatment of problems in various contexts and sets of conditions. The proof that any particular kinds of things are the right ones for a given context is then obtained by showing that they provide us with a good approximation to the essential features of reality in the context of interest. In other words, we require that theories formulated in terms of these abstractions lead to correct predictions, and to the control of natural processes in accordance with the plans that are made on the basis of these theories. When this does not happen, we must, of course, revise our abstractions until success is obtained in these efforts. Scientific research thus brings us through an unending series of such revisions in which we are led to conceptual abstractions of things that are relatively autonomous in progressively higher degrees of approximation, wider contexts, and broader sets of conditions.

8. THE PROCESS OF BECOMING

Thus far, we have been discussing the properties and qualities of things mainly in so far as they may be abstracted from the processes in which these things are always changing their properties and qualities and becoming other things. We shall now consider in more detail the characteristics of these processes which may be denoted by the general term of "motion". By "motion" we mean to include not only displacements of bodies through space, but also all possible changes and transformations of matter, internal, and external, qualitative and quantitative, etc.

Both the existence and the necessity for the process of motion described above have now been demonstrated in innumerable ways in all the sciences. Thus, the study of astronomy shows that the planets, stars, nebulae, and galaxies all take part in a very large number of kinds of characteristic motions. These motions follow from the effects of the gravitational forces which would start bodies moving even if they were initially at rest, and because of the inertia, which keeps them in motion. And as a result of these motions, over periods of time of the order of billions of years, new stars, new planets, new nebulae, new galaxies, new galaxies of galaxies, etc., can come into existence, while the older organization of things passes out of existence. On the earth, the science of geology has shown that the apparently permanent features of the surface are always changing. Thus, as a result of the flow of water and the action of wind, existing rocks and mountains, and even continents, are continually being worn away while subterranean motions are continually leading to the formation of new ones. The science of biology shows that life is a continual process of inexhaustible complexity in which various kinds of organisms come into being, live, and die. Indeed, every organism is maintained in existence by characteristic metabolic processes taking place within it, as well as by the motions necessary for it to obtain food and other materials from its environment. Over longer time, as a result of the effects of natural selection and other factors, the forms of life have had to evolve; and in this process, new species of organisms have come into existence while old species have died out. Over still longer periods of time, life itself has come into existence out of a basis of inanimate matter, very probably as a result of motions at the inorganic level of the kind suggested by Opharin*; and as conditions change it may later have to pass out of existence, perhaps to give way to something new, of which we can at present have no idea. In chemistry one sees that as a result of thermal agitation of the molecules and other causes, different chemical compounds must react to produce new kinds of compounds, while already existing kinds of compounds must be dissociated into simpler compounds. In physics we find, at the atomic level and below, a universal and ceaseless motion which follows as a necessary

* See Chapter I, Section 8.

consequence of the laws appropriate to these levels, and which is discovered to be more violent the deeper we penetrate into it. Thus, we have atomic motions, electronic and nucleonic motions, field motions, quantum fluctuations, probable fluctuations in a sub-quantum mechanical level, etc. Moreover, as happens at the higher levels, not only do the quantitative properties of things change in these motions (e.g. position, velocity, etc., of the various particles, the strength of the various fields, etc.), but so also do the basic qualities defining the modes of being of the entities, such as molecules, atoms, nucleons, mesons, etc., with which we deal in this theory.

In sum, then, no feature of anything has as yet been found which does not undergo necessary and characteristic motions. In other words, such motions are not inessential disturbances superimposed from outside on an otherwise statically existing kind of matter. Rather, they are inherent and indispensable to what matter is, so that it would in general not even make sense to discuss matter apart from the motions which are necessary to define its mode of existence.

Now, the various motions taking place in matter have the further very important characteristic that, in general, they are not and cannot be smoothly co-ordinated to produce simple and regular results. Rather, they are often quite complex and poorly co-ordinated and contain within them a great many relatively independent and contradictory tendencies.

There are two general reasons why such contradictory tendencies must develop; first because there are always chance disturbances arising from essentially independent causes, and secondly, because the systematic processes that are necessary for the very existence of the things under discussion are, as a rule, contradictory in some of their long-run effects. We shall give here a few examples taken from the fields that were discussed in the previous paragraph. Thus, in the field of astronomy, we find that partly as a result of chance disturbances from other galaxies and partly as a result of the laws of motion under the gravitational forces originating in the same galaxy, stars have a very complicated and irregular distribution of velocities going in all sorts of directions, etc., with the result that some systems of stars are being disrupted, while new systems are formed. On the surface of the earth, storms, earthquakes, etc., which are of chance origin relative to the life of a given individual, may produce conditions in which this individual cannot continue to exist; while a similar result can be brought about by old age, which follows from the effects of the very metabolic processes that are necessary to maintain life. Going on to the subject of physics, we see that both the effects of chance fluctuations and of the operation of systematic causal laws is continually leading to complicated and violent fluctuations in the various levels, which are not at all well co-ordinated with each other, and which quite often lead to contradictory tendencies in the motions. Indeed, these contradictory tendencies not only follow necessarily from the laws governing the motions, but must exist in order for many things to possess characteristic properties which help define what they are. For example, a gas would not have its typical properties if all the molecules had a strong tendency to move together in a co-ordinated way. More generally, the relative autonomy in the modes of being of different things implies a certain independence of these things, and this in turn implies that contradictions between these things can arise. For if things were co-ordinated in such a way that they could not come into contradiction with each other, they could not be really independent.

We conclude, then, that opposing and contradictory motions are the rule throughout the universe, and this is an essential aspect of the very mode of things.

Now it may be asked how it is possible for any kind of quality, property, entity, level, domain, etc., to have even an approximately autonomous existence, in the face of the fact that an infinity of relatively independent kinds of motions with contradictory tendencies are taking place in its environment and in its substructure. The answer is that the existence of any particular quality, property, entity, level, domain, etc., is made possible by a balancing of the processes that are tending to change it in various directions. Thus, in the simple case of a liquid,* we have a balancing of the effects of the inter-molecular forces tending to hold the molecules together, and the random thermal motions tending to disrupt the entire system. In a galaxy, we have a balancing of the gravitational forces against the centrifugal tendencies due to rotation and the disruptive effects of the random components of the motions of stars. In atoms we have a similar balancing of the attractive forces of the nucleus against the disruptive effects due to quantum fluctuations in the electronic motions and the centrifugal tendencies due to rotations of the electrons around the nucleus. With living beings, we have a much more subtle and complex system of balancing processes. The full analysis of this process naturally cannot yet be made. But already we can see that the two essential directions of processes in living beings are those leading to growth and those leading to decay. If the growth processes go unchecked, then a typical possible result is the development of a cancer, which eventually destroys the organism. On the other hand, if the opposite processes go unchecked, then the

* See Chapter II, Section 10.

organs will atrophy and wither away, and the organism will again eventually be destroyed. The maintenance of life then requires an approximate balancing of the destruction and decay of tissue by fresh growth.

Now it is clear that if qualities, properties, entities, domains, levels, etc., are maintained in existence by a balance of the processes tending to change them, then this balance can, in general, be only an approximate and conditional one. As a result, any given thing is subject to being changed with changing conditions, both by changes of conditions that are produced externally and by changes that may be necessary consequences of internal motions connected with the very mode of being of the thing in question. To illustrate this point, let us return to the problem of a liquid. As long as the temperature, pressure, etc., of the liquid are held constant, the balance of molecular processes that maintains the liquid state will be continued. But to think of an isolated specimen of a liquid is evidently an abstraction. Any real liquid exists in some kind of environment, which cannot fail to change with enough time. Thus, if the container is on the earth, it will be subject to changes of temperature, to storms and earthquakes that may destroy any temperature-stabilizing mechanism surrounding it, and over longer periods of time to geological processes that may have similar effects. Thus, it can safely be predicted that if, for example, we consider a period of a hundred million years, no particular specimen of a liquid will remain a liquid throughout the whole of this time. Analysing this problem further, we see that, as we consider broader contexts and longer periods of time, there will be more and more opportunities for conditions to change in such a way that any particular balance of processes is fundamentally altered. This is because it will be able to come into reciprocal relationships with more and more relatively autonomous entities, domains, systems, etc., the motions of which can come to influence the processes in question. Indeed, if we go to the extreme of considering supergalactic regions of space and corresponding epochs of time, we see that there is a possibility for such a broad range of changes of conditions that every kind of entity, domain, system, or level will eventually be subject to fundamental changes, even to destruction or extinction, while new kinds of entities, domains, and levels will come into existence in their place. For example, there is currently under discussion a theory in which it is assumed that some five billion years ago or more the parts of the universe that are now visible to us were originally concentrated in a comparatively small space having an extremely high temperature, and a density so high that neither atoms nor nuclei, nor electrons, nor protons, nor neutrons as we now know them could have existed. (Matter would then have taken some other form about which we cannot have much idea at the present.) This particular section of the universe is then assumed to have exploded, and subsequently to have cooled down to give rise ultimately to electrons, protons, neutrons, atoms, dust, clouds, galaxies, stars, planets, etc., by means of a series of processes into which we need not go further here. The recession of the stars, suggested by the so-called red shift,* would then be a residual effect of the velocities imparted to matter in this explosion. Now, it is very important to emphasize how speculative and provisional large parts of this theory are.† Nevertheless, for our purposes here, it is interesting in that it gives an example of how widespread could be the effects of a breaking of the balance of opposing processes within the previously existing highly dense state of matter; for the resulting explosion would have given rise to everything that exists in the part of the universe that is now visible to us.

In any case, whatever may have been the at present practically unknown earlier phase of the process of evolution of this particular part of the universe, there exists by now a considerable amount of evidence suggesting that the galaxies, the stars, and the earth come from some quite different previously existing state of things. With regard to what happened on our planet after it came into existence, we have of course much better evidence coming from traces left in the rocks, fossils, etc. Then, coming to the consideration of the

* The "red shift" of the spectral lines of stars has been interpreted as a Doppler shift due to a recessional motion. If this interpretation is correct, then the stars are receding from each other with a velocity that is more or less proportional to their distances. The most distant stars visible would have speeds as high as 10,000 miles a second, and still more distant stars would presumably have still higher velocities. However, there are many possible explanations for the same phenomenon; e.g. perhaps the behaviour of light over long distances is slightly different from that predicted by Maxwell's equations, in such a way that the frequency of light diminishes as it is transmitted through space.

† In the actually published forms of this theory, it is assumed that the *whole universe* (and not just a part of it) was originally concentrated into the small space referred to above. Even if we do not make this additional assumption, the theory is already quite speculative. But this additional assumption is based on Einstein's theory of general relativity, which has been proved to a rather low level of approximation only in weak gravitational fields for low concentrations of matter and over limited regions of space. A gigantic extrapolation is then made to gravitational fields of fantastic intensity, to unheard-of concentrations of matter, and to a region of space that includes nothing less than the whole universe. While this extrapolation cannot be proved to be wrong at present, it is in any case an example of extreme mechanism. If we divest the theory of these irrelevant and unfounded extrapolations, then the hypothesis is still, however, interesting to consider.

origin of life, we have the hypothesis of Opharin,* which gives at least the general outlines of how living matter could have come into existence on the earth. Here we see the importance of the incomplete co-ordination and contradictory character of the various kinds of processes that took place on the earth at the time in question; for storms, ocean currents, air currents, etc., would have led to a chance mixing of various organic compounds until at last a substance appeared that began to reproduce itself at the expense of the surrounding organic material. As a result, the contradictory character of the motions at the inorganic level created the conditions in which a whole new level could come into existence, the level of living matter. And from here on, changes in the inanimate environment ceased to be the only causes of development. For a fundamental property of life is that the very processes that are necessary for its existence will change it. Thus, in the case of the individual living being, the balance of growth and decay is never perfect, so that in the earlier phases of its life, the organism grows, then it reaches approximate balance at maturity, and then the processes of decay begin to win out, leading to death. With regard to the various species of living beings considered collectively, these provide each other with a mutual environment, both through their competition and through their co-operation. Thus as a result of the very development of many kinds of living beings, the environment is changed in such a way that the balance of the processes maintaining the heredity of such species is altered, and the result is the well-known evolution of the species.

In sum, then, we see that the very nature of the world is such that it contains an enormous diversity of semi-autonomous and conflicting motions, trends, and processes. Thus, if we consider any particular thing, either the motions taking place externally to it or those taking place internally and which are inherent aspects of its mode of being will eventually alter or destroy the balance of processes that is necessary to maintain that thing in existence in its present form and with its present characteristics. For this reason, any given thing or aspect of that thing must necessarily be subjected to fundamental modifications and eventually to destruction or decay, to be replaced by new kinds of things.

In conclusion, the notion of the qualitative infinity of nature leads us to regard the eternal but ever-changing process of motion and development described above as an inherent and essential aspect of what matter is. In this process there is no limit to the new kinds of things that can come into being, and no limit to the number of kinds of transformations, both qualitative and quantitative, that

* See Chapter I, Section 8.

can occur. This process, in which exist infinitely varied types of natural laws, is just the process of *becoming*, first described by Heraclitus several thousand years ago (although, of course, by now we have a much more precise and accurate idea of the nature of this process than the ancient Greeks could have had).

9. ON THE ABSTRACT CHARACTER OF THE NOTION OF DEFINITE AND UNVARYING MODES OF BEING

It is clear from the preceding section that the empirical evidence available thus far shows that nothing has yet been discovered which has a mode of being that remains eternally defined in any given way. Rather, every element, however fundamental it may seem to be, has always been found under suitable conditions to change even in its basic qualities, and to become something else. Moreover, as we have also seen, the notion of the qualitative infinity of nature implies that every kind of thing not only can change in this very fundamental way but that, given enough time, conditions in its infinite background and substructure will alter so much that it *must* do so. Hence, the notion of something with an exhaustively specifiable and unvarying mode of being can be only an approximation and an abstraction from the infinite complexity of the changes taking place in the real process of becoming. Such an approximation and abstraction will be applicable only for periods of time short enough so that no significant changes can take place in the basic properties and qualities defining the modes of being of the things under consideration.

When we come to times that are long enough for the basic kinds of things entering into any specific theory to undergo fundamental qualitative changes, then what breaks down is the assumption that we can specify the modes of being of these things *precisely* and *exhaustively* in terms of the concepts that were applicable before this change took place. Indeed, the very fact that a thing is able to undergo a qualitative change is itself a property that is an essential part of the mode of being of the thing and yet a property that is not contained in the original concept of it. For example, as we saw in Chapter I, Section 6, the fact that the liquid, water, turns into steam when heated and ice when cooled, is a basic property of the liquid in question, without which it could not be water as we know it. Nevertheless, the original concept of water as nothing more than a liquid evidently does not contain these possibilities, either explicitly or implicitly, as necessary properties of this liquid. Hence, this concept does not give a precise and exhaustive representation of all the properties of the liquid in question.

Now the way one usually deals with this problem is to regard the transformations between solid, liquid, and vapour that take place at certain temperatures as part of the qualities defining the mode of being of a single broader category of substance; viz. water. But now the same kind of problem arises again at a new level. For the laws governing the transformations of these qualities are, in turn, being regarded as part of an eternal and exhaustive specification of the properties of the substance, water. On the other hand, in reality this law is applicable and has meaning only under limited conditions. For example, it will no longer have relevance at temperatures and denotes of matter so high that there can be no such things as atoms, and therefore no such a substance as water. Thus, we are led to include water as a special state of a still broader category of things (e.g. systems of electrons, protons, neutrons, etc.) and the laws governing the transformation of water into other kinds of substances as a part of the mode of being of this still broader category. But if *all* things eventually undergo qualitative transformations, then the process described above will never end. Thus we conclude that the notion that all things can become other kinds of things implies that a complete and eternally applicable definition of any given thing is not possible in terms of any finite number of qualities and properties.

If, however, we now start from the opposite side, viz. from the notion of the qualitative infinity of nature, we are then immediately able to arrive at a type of definition of the mode of being of any given kind of thing that does not contradict the possibility of its becoming something else. For, as we saw in Section 7, the reciprocal relationships between all things then imply that no given thing can be *exactly* and *in all respects* the kind of thing that is defined by any specified conceptual abstraction. Instead, it is always *something more* than this and, at least in some respects, *something different*. Hence, if the thing becomes something else, no unresolvable contradiction is now necessarily implied. For it is in any case never exactly represented by our original concept of it. Logically speaking, what this point of view towards the meaning of our conceptual abstractions does is, therefore, to create room for the possibility of qualitative change, by leading us to recognize that those aspects of things that have been ignored may, under suitable conditions, cease to have negligible effects, and indeed may become so important that they can bring about fundamental changes in the basic properties of the things under consideration.

We may illustrate the above conclusions by returning to a more detailed discussion of the transformations between steam, liquid water, and ice. Thus, as we saw in Chapter II and in the present chapter, the macroscopic concept of a certain state of matter (e.g.

gaseous, liquid, or solid) leaves out of account an enormous number of kinds of factors that are not and cannot be defined in the macroscopic domain alone. Among these are the motions of the molecules constituting the fluid quantum fluctuations, field fluctuations, nuclear motions, mesonic motions, motions in a possible subquantum mechanical level, and so on. In short, we may say that the real fluid is enormously richer in qualities and properties than is our macroscopic concept of it. It is richer, however, in just such a way that' these additional characteristics may, in a wide variety of applications, be ignored in the macroscopic domain. Nevertheless, when we come to the problem of understanding why transformations between gas, liquid, and solid are possible, we can no longer completely ignore the additional properties of the real fluid. Thus, as shown in Chapter II, Section 10, the molecular motions are able to explain at least the essential features of the transformation in the system from a state in which one set of qualities (i.e. those corresponding to a gas) are the determinant, dominant, and controlling factors to a state in which these are replaced by another set of qualities (e.g. those corresponding to a liquid). Moreover, according to the notion of the qualitative infinity of nature, the same general kind of result is obtained for all things, including, for example, even the most fundamental entities that may have been discovered at any particular stage in the development of physics.

Not only is the notion of unvarying and exhaustively specifiable modes of being of things an abstraction that fails for periods of time that are too long (because of the possibility of fundamental qualitative changes), but it also fails for times that are too short. This is because the characteristic properties and qualities of a thing depend in an essential way on processes that are taking place in the background and substructure of the thing in question. Thus, for example, the properties of an atom (e.g. spectral frequencies, chemical reactivity, etc.) arise and are determined mainly in the process of motion of the electrons in the orbit, which take a period of time of the order of 10^{-15} seconds. Over shorter periods of time, however, the properties of an atom as a whole are so poorly defined that it is not even appropriate to consider them as such. A better conception of what the atom is can then be obtained by regarding it as a collection of electrons in motion around the nucleus. But as we shorten the period of time still further, the same problem arises with regard to electrons, protons, neutrons, mesons, etc. And if we go to a larger scale, the reader will readily see that a similar behaviour is obtained (e.g. the existence of a living being is maintained by inner metabolic and nervous processes that are fast in comparison with the period in which it makes sense to define the basic characteristics of such a being). Indeed, the notion of the qualitative infinity of nature implies that such behaviour is inevitable. For, as we saw in the previous section, each kind of thing is maintained in existence by a balance of opposing processes in its infinite background and substructure, which are tending to change it in different ways. Thus, the properties of such a thing can be defined only over periods of time long enough so that the average of the effects of all these processes does not fluctuate significantly.

It is clear, then, that all our concepts are, in a great many ways, abstract representations of matter in the process of becoming. The choice of such abstractions is, however, limited by the requirement that they shall represent what is essential in a certain context to a suitable degree of approximation and under appropriate conditions.

The particular kind of abstraction that is used may evidently then vary, depending on what the context is. Thus, in theories of simple types of phenomena where things can be approximated as being in equilibrium, the modes of being of the basic entities and properties may be conceived of as completely static (e.g. as in statics and in thermodynamics). In the study of phenomena where motion is important, however, a higher level of abstraction is needed. For example, in mechanics one considers a system of particles which can change their positions without ceasing to be particles. In other words, the being of the particles is *indifferent* to their positions, and we can therefore consider them to be in motion through space. But the unvarying *laws* applying to these motions are now regarded as constituting an essential part of the modes of being of the particles in question. Thus, we have not escaped the necessity for considering unvarying and exhaustively specifiable modes of being. Of course, we could in principle go further and suppose, for example, that even the laws of motion of the particles were evolving with time. But then we would still be assuming that the higher laws applying to this process of evolution were themselves unvarying and in principle exhaustively specifiable in their form. On the other hand, according to the notion that everything takes part in the process of becoming, even these latter features of the laws could not ever really be completely unvarying and exhaustively specifiable in terms of a finite number of kinds of things.

We conclude, then, that we must finally reach a stage in every theory where we introduce the notion of something with unvarying and exhaustively specifiable modes of being, if only because we cannot possibly take into account all the inexhaustibly rich properties, qualities, and relationships that exist in the process of becoming. At this point, then, we are making an abstraction from the real process of becoming. Whether the abstraction is adequate or not depends on whether or not the specific phenomena that we are studying depend significantly on what we have left out. With the further progress of science, we are then led through a series of such abstractions, which furnish ever better representations of more and more aspects of matter in the concrete and real process of becoming.

Now, when we refer to the process of becoming by the word "concrete", we mean by this to call attention to the quality of being special, peculiar, and unique that one always finds to be characteristic of real things when one studies them in sufficient detail. For example, if we consider any concept (e.g. apples), then this concept contains nothing in it that would permit us to distinguish one apple from another. We may then indicate other qualities which make such a distinction possible (e.g. red apples, hard apples, sweet apples, etc.). Evidently, no finite number of such qualities can ever give a complete representation of any specific example of a real apple. Of course, by going deeper (e.g. by giving the physical and chemical state of each part of the apple) we could come closer to our goal. But this process could never end. For even the modes of being of the individual atoms, electrons, protons, etc., inside the apple are in turn determined by an infinity of complex processes in their substructures and backgrounds. Thus, we see that because every kind of thing is defined only through an inexhaustible set of qualities each having a certain degree of relative autonomy, such a thing can and indeed must be *unique*; i.e. not completely identical with any other thing in the universe, however simila ther two things may be.*

Carrying the analysis further, we now note that because all of the infinity of factors determining what any given thing is are always changing with time, *no such a thing can even remain identical with itself* as time passes. In certain respects, this brings us to a deeper notion of the process of becoming than we had before. For at each instant of time, each thing has, when viewed from one side, an enormous (in fact infinite) number of aspects which are in common with those that it had a short time ago. Indeed, if this were not so, it would not be a thing; i.e. it would not preserve any kind of identity

* According to the Pauli exclusion principle, any two electrons are said to be "identical". This conclusion follows from the fact that within the framework of the current quantum theory there can be no property by which they could be distinguished. On the other hand, the conclusion that they are *completely* identical in *all* respects follows only if we accept the assumption of the usual interpretation of the quantum theory that the present general form of the theory will persist in every domain that will ever be investigated. If we do not make this assumption, then it is evidently always possible to suppose that distinctions between electrons can arise at deeper levels.

at all. On the other hand, when viewed from another side, it has an equally enormous (in fact infinite) number of aspects that are not those that it had a short time ago. For typical sorts of things with which we commonly deal, however, these latter aspects are not essential in the normal contexts and conditions with which we work. In new contexts (e.g. a sub-atomic or a super-galactic time scale) or under new conditions (e.g. very high temperatures), these aspects may, however, take on a crucial importance.

We are in this way led to the conclusion that the process of becoming will necessarily have, at each moment, certain aspects that are concrete and unique. In other words, each thing in each moment of its existence must have certain qualities which, in some respects, belong uniquely to that thing and to that moment. The notion of unvarying and exhaustively specifiable modes of being is then an abstraction obtained, in general, by considering what is common to the same thing at different moments, or to many similar things at the same moment. In doing this, we evidently ignore the differences between these things, which are just as essential a side of them as are their similarities. By abstracting in more detail from these differences, we are then led to see newer but subtler aspects in which these differences contain common or similar relationships that apply to all of these things. Thus, the uniqueness of each thing at each instant of time is reflected in our abstract concepts by the limitless richness and complexity of the concepts that one needs to obtain a better and better abstract representation of matter in the process of becoming, or, in other words, by the inexhaustibility of the qualities that are to be found in nature.

10. REASONS FOR INADEQUACY OF LAPLACIAN DETERMINISM

We are now ready to see why the mechanistic determinism of Laplace does not apply if the notion of the qualitative infinity of nature is correct. For this kind of determinism implies that the laws of nature are such as to permit the super-being of Laplace to know them in their totality. On the other hand, according to the point of view that we have been presenting, this is impossible.

First of all, let us recall that no matter how far one goes in the expression of the laws of nature, the results will always depend in an unavoidable way on essentially independent contingencies which exist outside the context under investigation, and which are therefore undergoing chance fluctuations relative to the motions inside the context in question. For this reason, the causal laws applying inside any specified context will evidently not be adequate for the perfect prediction even of what goes on inside this context alone.

Secondly, however, the essential independence of different contexts implies that the processes taking place within a given context cannot provide a complete and perfect reflection of what goes on in the infinite totality of possible contexts. For example, because of the cancellation of chance fluctuations, the precise details of atomic motions are not usually reflected to any significant extent in the laws of the macroscopic level. The laws of each new context must then, in general, be discovered with the aid of new kinds of experiments, set up so as to create conditions in which the laws of the new context under investigation are significantly reflected in the behaviour of the apparatus. Hence, even to know what the totality of all the laws of nature is, the super-being would have to do an infinity of different kinds of experiments, each of which would give results that depended significantly on the laws of a different context, so that he could thereby obtain the necessary information. In doing this, he would have to be able to discover not only all the already operating kinds of laws, but also all the new laws that are expressible only in terms of the infinity of new qualities, new entities, and new levels that are going to come into being, all the way into the infinite future. It is evident, then, that if the Laplacian super-being resembles us to the extent of obtaining his knowledge through a series of investigations of partial segments of the universe, and not, for example, by Divine revelation or by *a priori* intuitions which he finds by plumbing the depths of his own mind, he will never be able to predict the entire future of the universe or even to approach such a prediction as a limit, no matter how good a calculator he may be. And if he did have such revelations or intuitions, a calculation would hardly be necessary, since the detailed prediction of the behaviour of the universe would then require a miracle only slightly greater than that by which he would learn the basic laws of the universe in the first place.

We see, then, that the behaviour of the world is not perfectly determined by any possible purely mechanical or purely quantitative line of causal connection. This does not mean, however, that it is arbitrary. For if we take any given effect, we can always in principle trace it to the causes from which its essential aspects came. Only as we go further and further back into the past, we discover three important points: viz. first, that the number of causes which contribute significantly to a given effect increases without limit; secondly that more and more qualitatively different kinds of causal factors are found to be significant; and finally, that these causes depend on new contingencies leading to new kinds of chance. For example, let us consider an eclipse of the moon. Over moderate periods of time this is a fairly precisely predictable event, which is determined mainly by the co-ordinates and momenta of the earth and the moon relative to the sun. But the longer the time that we consider, the more precise this determination must be, in order to make possible a prediction of the effect with a given accuracy. For the details of the motion become very sensitive to the precise initial conditions. As a result, perturbations arising from other planets, from tides in the earth, the moon, the sun, and still other essentially independent contingencies become significant. Over long enough periods of time, even the fluctuations arising from the molecular motions could in principle come to have significant effects; but before this could really become important, we should have gone so far into the past as to reach the qualitatively different phase of the gaseous nebulae from which the earth, moon, and sun came. Here we see that the random motions of the gas molecules in these nebulae contributed to making the eclipse eventually occur in the way that it did. If we go further back, we might reach the dense state of matter, that perhaps existed before the explosion that may have led to the present state of the part of the universe that is now visible to us. Then, the motions of the entities existing in this previous state, whatever they may have been, would have contributed to making the eclipse occur in the way that it did. But these motions would be contingent on something still earlier. And so on without limit. It is clear, moreover, that the eclipse of the moon is a phenomenon that is subject to an exceptionally simple type of determination, because of the approximate isolation of the earth and moon from other things. In other processes, where the degree of isolation is much less, the intertwining and fusion of the effects of more and more contingencies and more different qualities as we go further back is much greater. Thus, over an infinite period of time, the determination of even the essential features of an effect is evidently not purely mechanical, because it involves not only an infinite number of contingent factors but also an infinity of kinds of qualities, properties, laws of connection, all of which themselves undergo fundamental changes with the passage of time.

11. REVERSIBILITY VERSUS IRREVERSIBILITY OF THE LAWS OF NATURE

In this section we shall make a few remarks concerning the implications of the qualitative infinity of nature with regard to the question of whether the laws of nature are reversible or irreversible.

It is well known that thus far the laws of microscopic physics have demonstrated themselves to be reversible. This follows from the fact that starting with any solution of the basic equations for the system (Newton's laws of motion, the laws of relativity, the laws of quantum theory), another possible solution can be found by replacing the time,* t, by its negative, $-t$. Physically this means that given any motion, it is always possible, in principle at least, for a similar motion to take place, which is, however, executed in the reverse order. Of course, to obtain such a reversal of motion in reality, we would have to alter the boundary conditions appropriately (e.g. reverse all the velocities of the various particles, rates of change of the fields, etc.). Such a reversal does not, in general, occur spontaneously, at least within any practically significant periods of time. To show that this is so let us consider, for example, two boxes of gas, one containing hydrogen and the other containing oxygen, and let us imagine that we open a tube that connects them. As is well known, the gases will diffuse into each other. The reason is, of course, that the complicated and irregular motions of the hydrogen molecules will tend to carry them into the chamber originally containing oxygen, while similar motions of the oxygen molecules will tend to carry them into the chamber originally containing the hydrogen. As we have seen in Chapter II, Section 12, such processes can be treated in terms of the laws of chance, so that the theory of probability can be applied to them. Since over a long period of time it is equally probable that any particular molecule will occupy any given region of space, we conclude that on the average and in the long run we will obtain a practically uniform mixture of hydrogen and oxygen. It is characteristic of the laws of chance, however, that fluctuations away from the average can occur, although large fluctuations are very rare. A simple calculation, using the appropriate law of probability for these fluctuations, shows, for example, that a chance combination of motions that led all the hydrogen and oxygen back into their original containers would, under typical conditions, not occur for $10^{10^{10}}$ years (i.e. 1 followed by ten thousand million zeros). Clearly, then, although the motion may in principle reverse, the probability that this will happen is so small that we may for practical purposes ignore this possibility, especially considering the fact that, in any case, the containers of gas could not possibly last for such a long time.

It is possible by means of analysis described above to understand the observed irreversibility in various physical phenomenon, such as the flow of heat, the establishment of thermal and mechanical equilibrium in fluids, etc. But this still leaves us with a disturbing problem. For the above reduction of the observed irreversibility of

* In the case of the quantum theory, we must also replace the wave function, ψ, by its complex conjugate, but this does not change any probabilities of physical processes, which depend only on $|\psi|$.

certain large-scale phenomena to the effects of chance does not alter the fact that the fundamental equations of motion are reversible, so that there is no inherent reason why processes in general must necessarily always take place in one direction only, since either direction would in principle be possible. Thus, if all the velocities and rates of change of fields did actually manage to be reversed for any reason whatever (e.g. by chance), then heat could go from a lower to a higher temperature, water could flow from the sea back to its sources in the mountains, etc. The fact that these events are so fantastically improbable does not detract from the problem of principle presented here, which is this: "Do the generally irrevocable effects of the passage of time in so wide a range of fields really come out of nothing more than the random mixing or shuffling according to the laws of chance of molecular and other types of motion, the reversal of which is in principle possible but in practice too improbable to be considered as having any real importance?"

If we take into account the character of the laws of physics implied by the qualitative infinity of nature, however, we can immediately answer this question in the negative. For, as we have seen, the notion of a law that gives a perfect one-to-one mathematical correspondence between well-defined variables in the past and in the future, is only an abstraction, good enough to describe limited domains of phenomena for limited periods of time, but, nevertheless, not valid for all possible domains over an infinite time. Thus, as has been pointed out in Section 8, the very entities with which physics now works, satisfying the currently studied laws of physics, must have come into being at some time in the past, while changing conditions, brought about in part by the effects of just these laws, and in part by chance contingencies, will eventually lead to a stage of the universe in which new kinds of entities satisfying new kinds of laws will come into being. On a smaller scale, we see also that new levels, such as that of living matter, have come into being, in which characteristic new qualities and new laws appear. Thus, we are not justified in making unlimited extrapolations of any specific set of laws to all possible domains and over infinite periods of time. This means that the description of the laws of nature as in principle completely reversible is merely a consequence of an excessively simple representation of reality. When we consider the mechanical laws in their proper contexts of ever-changing basic qualities, it becomes clear that irrevocable qualitative changes do take place, which could not even in principle be reversed. This is because, for systems of appreciable complexity, the fundamental character of the laws that apply cannot be completely separated from the historical processes in which these systems come to obtain their characteristic properties.* The possibility of such a behaviour is especially clear with regard to living matter, for here the very mode of being an organism and the basic qualities and laws which define this mode of being arise in the process by which the organism comes into existence, and passes through the various stages of its life. Thus, it is quite impossible that a human being could become a human being except by a process of growth, through embryo, childhood, adulthood, etc. But when one analyses processes taking place in inanimate matter over long enough periods of time, one finds a similar behaviour. Only here the process is so much slower that the abstraction in which we conceive of matter as having properties that are independent of its specific historical development is usually quite good as long as one considers periods of time which are measured in units smaller than billions of years.

The importance of considering the impact of qualitative changes on the basic modes of being of things is also seen when we consider the predictions of the "heat death" of the universe, which were especially common towards the end of the nineteenth century. The "heat death" refers to the prediction that eventually, because of random mixing and shuffling of molecules, the temperature of the universe would become uniform, and therefore, at least on the large scale, nothing could happen, so that the universe would be "dead". However, long before this comes about, it is evidently quite possible and indeed very likely that qualitatively new developments reflecting the inexhaustible and infinite character of the universal process of becoming will have invalidated predictions of the type described above. For example, just as there may have been a time before molecules, atoms, electrons, and protons existed, the further evolution of the universe could also lead to a new time in which they cease to exist, and are replaced by something else again. And new sources of energy coming from the infinite process of becoming may be made available even if atoms, molecules, etc., continue to exist. Thus, in the last century only mechanical, chemical, thermal electrical, luminous, and gravitational energy were known. Now we know of nuclear energy, which constitutes a much larger reservoir. But the infinite substructure of matter very probably contains energies that are as far beyond nuclear energies as nuclear energies are beyond chemical energies. Indeed, there is already some evidence in favour of this idea. Thus, if one computes the "zero point" energy due to quantum-mechanical fluctuations in even one cubic centimetre of space, one comes out with something of the order of 10^{38}

* Of course, this may not be the only reason or even the main reason for the observed irreversibility in nature, but in any case, for this reason alone, irreversibility would follow.

ergs, which is equal to that which would be liberated by the fission of about 10^{10} tons of uranium.* Of course, this energy provides a constant background that is not available at our level under present conditions. But as the conditions in the universe change, a part of it might be made available at our level.

Not only is the qualitatively and quantitatively infinite universal process of becoming too complex even to reverse itself or to come to some kind of final equilibrism, but it also cannot go in a cycle. For even if the laws applying in certain contexts and conditions should be consistent with a cyclical universe, such laws will always leave out an infinity of new kinds of factors, which will in the long run become important as conditions change sufficiently. Unless these new factors are exactly coordinated with those already existing in more limited contexts and sets of conditions, they will eventually break the cycle and bring in fundamental qualitative changes. But because of their relative and approximate autonomy, these factors would not in general be coordinated in such a way. Hence a cyclical behaviour would also be inconsistent with the character of the universe that we have been considering here.

In conclusion, then, the notion of the qualitative infinity of nature implies that the development of the universe in time will lead to an inexhaustible diversity of new things.

12. ABSOLUTE VERSUS RELATIVE TRUTH— THE NATURE OF OBJECTIVE REALITY

We shall now sum up the ideas developed in this chapter, and indeed throughout the whole book in terms of a treatment of the implications of the nature of the qualitative infinity of nature, with regard to the problems of the absolute *v.* the relative character of truth, and of what, in the framework of this point of view, is meant by the concept of objective reality.

To begin with, let us recall that we are led to understand nature in terms of an inexhaustible diversity and multiplicity of things, all of them reciprocally related and all of them necessarily taking part in the process of becoming, in which exist an unlimited number of relatively autonomous and contradictory kinds of motions. As a result no particular kind of thing can be more than an abstraction

---

* Actually, according to present theories, this energy is infinite, but if one assumes that the theory is valid down to fluctuations having wave-lengths of the order of 10^{-13} cm., then the above value of the energy is obtained. This wave-length was chosen, because it is generally believed that current theories of quantum electro-dynamics break down for shorter wave-lengths, and break down in such a way that the effects of quantum-fluctuations become finite. Thus, in a very rough estimate, we may ignore the effects of wave-lengths shorter than 10^{-13} cm.

from this process, an abstraction that is valid within a certain degree of approximation, in definite ranges of conditions, within a limited context, and over a characteristic period of time. Such an abstraction evidently cannot represent an absolute truth; for to do this it would have to be valid without approximation, unconditionally, in all possible contexts, and for all time. Hence, any particular theory will constitute an approximate, conditional, and relative truth.

We may then ask the question, "Does the fact that any given theory can only be approximately, conditionally, and relatively true mean that there is no objective reality? To see that this is not so, it is only necessary to ask the further question of whether the behaviour of things is arbitrary. For example, would it be possible for us to choose the natural laws holding within a given degree of approximation and in a particular set of conditions at will, in accordance with our tastes, or with what we feel would be helpful for us in the solution of various kinds of practical problems? The fact that we cannot actually do this shows that these laws have an objective content, in the sense that they represent some kind of necessity that is independent of our wills and of the way in which we think about things. This does not mean that we cannot, in general, make our own choices as to what we will or will not do. But unless these choices are guided by concepts that correctly reflect the necessary relationships that exist in nature, the consequences of our actions will not in general be what we chose, but rather something different, and something that is quite often what we would have chosen not to aim for if only we had known what was really going to come out of our actions.*

It is true, of course, that the same natural laws can often be treated with the aid of a series of very different kinds of conceptual abstractions. Thus, in the domain of classical physics, we could equally well work with the abstractions that are appropriate to classical mechanics (e.g. particles following orbits defined by definite laws), or we could utilize those that are appropriate to quantum mechanics (e.g. systems existing in discrete states, to which apply laws of prob-

* For example, we could choose to step out of a window and fly upwards into the sky. If we tried to do this, however, we would fall downwards. The same thing would happen even if we had been guided in our actions by a set of concepts which led us to the conclusion that it was possible to fly upwards merely by flapping one's arms and saying certain magic words. Actually, if we wish to fly, what we must do is to have a deeper and more accurate conception of the laws of dynamics; and on the basis of this to construct suitable devices such as aeroplanes, dirigibles, rockets, etc. Thus, in the last analysis, the laws of nature do not depend on how we think about it or on what we choose to do, but our actions must be guided by correct conceptions of these laws if they are to lead to the results that we aim for.

ability), and then take the limit of large numbers of very small quanta. Which of these very different procedures we used would not matter in this particular domain, because both would lead to essentially the same results. Indeed, as we pointed out in Chapter I, Section 10, the different possible conceptual abstractions here play the rôle of various views of different aspects of the same basic reality. *To the extent that these different abstractions have a common domain of validity,* they must lead to the same consequences (just as different views must be consistent with each other in their domain of overlap).

It is clear from the preceding discussion that a necessary part of the definition of the extent to which a given law is true lies in the delimitation of its *domain of validity*. To accomplish the definition of this domain, we must find the *errors* in the law in question. For the more we know about these errors the better we will know the conditions, context, and degree of approximation within which this law can correctly be applied and therefore the better we will know its domain of validity.

Now, if there were a final and exhaustively specifiable set of laws which constituted an absolute truth, we could regard all errors as purely subjective characteristics, resulting from uncertainty in our knowledge concerning this absolute truth. On the other hand, in terms of the notion of the qualitative infinity of nature, we see that *every* law that can possibly be formulated has to have errors, simply because it represents nature in terms of some finite set of concepts, that inevitably fail to take into account an infinity of additional potentially or actually significant qualities and properties of matter. In other words, associated with any given law there must be errors that are essential and objective features of that law resulting from the multitudes of diverse factors that the law in question must neglect.* Thus each law inevitably has its errors, and these are just as necessary a part of the definition of its true significance as are those of its consequences that are correct.

It is clear from the above discussion that scientific research does not and can not lead to a knowledge of nature that is completely free from error. Rather it leads and is able to lead only to an unending process in which the degree of truth in our knowledge is continually increasing. The first step in any part of this process is generally accomplished with the aid of new kinds of experiments and observa-

---

* In Section 6 we saw a special case of the essential character of errors in the laws of nature. Thus, casual laws inevitably contain errors resulting from the neglect of chance fluctuations originating in contexts external to what is treated by the casual law in question. Vice versa, laws of chance must contain errors resulting from the neglect of casual interconnections brought about by the laws operating in broader contexts.

tions or with more accurate forms of already familiar kinds of experiments and observation, which serve to disclose some of the errors that are inevitably present at any particular stage in the development of our theories. The next step, then, comes after we have discovered some of the new laws that apply in the newer and broader domains to which we have in this way been led. For, as we have seen in terms of a number of examples given in previous chapters,* these new laws not only approach the older laws as approximations holding in limiting cases, but they also help to specify the degree of approximation and the conditions within which the older laws will actually hold. Thus, with the further progress of science into new domains, it becomes possible for us to define the errors in older laws in more and more detail and in more and more respects, and in this way to delimit the domains of validity of these laws more precisely and more nearly completely.

Now, if we could determine *all* the errors in a given law perfectly, we would know the absolute truth about that law. For we would then know just when and where, and to just what degree of approximation, it is valid, so that we would *never* be led to wrong predictions through using it. Of course, we cannot determine all the errors in any given law completely; and as a result, we can never actually reach such an absolute truth with regard to the law in question. Nevertheless, in many fields we are able to determine the errors so well in this way that we can say that at least for the specific domains under consideration, we are approaching closer and closer to an absolute truth (more or less as we are able to come closer and closer to a representation of a curved figure by inscribing it in a series of polygons with more and more sides and thus to give a series of successive approximations that converges towards a definite and limited result).

With regard to nature as a whole, however, it cannot be said that this continual process of disclosure of errors in our theories is leading us through a series of successive approximations that converges on some fixed and limited goal, which constitutes an absolute truth. For as science progresses, we find that the process of uncovering the errors in previous theories continually points towards the existence of more and more new kinds of things, which were not significant in contexts and conditions studied up to a certain point in the development of our researches, but which may be of crucial importance in new contexts and conditions. As a result, the goal of an absolute truth that applies in all possible contexts and conditions keeps on receding beyond the new horizons that appear before us

* E.g. the approach of quantum theory to classical theory and of relativity to Newtonian mechanics.

as we continue our studies of the inexhaustible characteristics of nature in more and more detail and in more and more different ways. It is true that there is nothing in the structure of the universe that could prevent us from eventually coming in these studies to know about any given thing. Indeed, as our understanding of the reciprocal relationships between things grows better, we will be able to make more and more kinds of measurements which probe deeper and deeper into the structure of the universe and which reach out further and further from the particular region of space and time in which our existence is centred. For these relationships will enable us to infer the character of things that are on different levels or far away from us, on the basis of experiments and observations on things that are on our level and which are in the domain of the space and time that is immediately accessible to us. Thus, any *given* kind of thing is, in principle, knowable. On the other hand, no matter how far even the whole of humanity may progress in any specified period of time, however long, it cannot reach or even approach a complete, perfect, and unconditional knowledge of reality as a whole. Thus, with regard to reality as a whole, the analogy of the approach to a given curved figure by means of a set of smaller and smaller tangent lines is not appropriate. A better analogy would be a particle in Brownian motion, the path of which can be approached in this way only within a certain degree of approximation, but which must be treated, as we go deeper and deeper, with the aid of more and more new qualities and properties, such as those associated with the atoms and molecules in motion, quantum fluctuations, etc.

If we stopped at this point in our analysis of the problem of truth, however, we would be focusing our attention on the side of the infinite diversity and multiplicity of things in the universe, and thus we would lose sight of how they are united as different aspects of one world. For we would tend to think of things and qualities as strung out one after the other in a never-ending line or as strewn through space in a limitless chaos. In order to see the world from the side of its being a unity, we must start from the notion that the basic reality is the totality of actually existing matter in the process of becoming. It is the basic reality because it has an independent kind of existence such that none of its characteristics depend on anything else that is outside of itself. This is so because the *totality* of matter in the process of becoming contains, by definition, everything that exists. If we find that something is outside of any given part of what we are considering, this merely means that we must define a broader category, which includes the part in question as well as what is outside of it. Thus, even though the existence and the characteristic defining the mode of being of any given thing can, and indeed must, be contingent on other things, that of the infinite totality of matter in the process of becoming cannot, because whatever it might be contingent on is also by definition contained in this totality.

We then come to the question of defining in detail what *is* this totality of matter in the process of becoming. By this we mean that we wish to specify its basic properties and qualities, and to delimit its general characteristics.

Now, the most essential and fundamental characteristic of the totality of matter in the process of becoming lies precisely in the fact that it can be represented only with the aid of an inexhaustible series of abstractions from it, each abstraction having only an approximate validity, in limited contexts and conditions, and over periods of time that are neither too short nor too long. These abstractions have many rationally understandable relationships between them. Thus, they represent things that stand in reciprocal relationships with each other, and each theory, expressed in terms of a specific kind of abstraction, helps to define the domains of validity of different theories, expressed in terms of other kinds of abstractions. The fact that all these relationships exist is not surprising, since every theory is, in any case, some kind of abstraction from the same totality of matter in the process of becoming. Vice versa, the fact that we need an inexhaustible series of such abstractions for the better and better representation of reality as a whole is also not surprising, provided that we recall that, as we saw in Section 9, this reality is concrete; i.e. has aspects that are unique for each thing in each amount of its existence.

The definition of the concrete characteristics of the totality of matter in the process of becoming can then be accomplished in unlimited detail in terms of relationships among the things that one can abstract out of this process itself. For each thing that exists in this process can be defined, to successively better approximations and in progressively wider contexts, in terms of its reciprocal relationships with more and more other things. This is the basic reason why the study of any one thing throws light on other things, and thus eventually leads back to a deeper understanding of its own properties. In fact, if it were possible to define the totality of all reciprocal relationships between things, this would enable us to define matter in the process of becoming completely. For every thing that exists, including all its characteristic properties and qualities, every event that happens, and every law relating these events and things, is defined only through such reciprocal relationships. And what more can there be to define about matter in the process of becoming, except that which does not exist, has no properties and qualities, satisfies no laws, does not happen, and which is therefore precisely

169

nothing? Of course, as has already been pointed out, we cannot actually come to know all these reciprocal relationships in any finite time, however long. Nevertheless, the more we learn about them, the more we will know about what matter in the process of becoming is, since its totality is defined by nothing more than the totality of all such relationships.

In conclusion, a consistent conception of what we mean by the absolute side of nature can be obtained if we start by considering the infinite totality of matter in the process of becoming as the basic reality. This totality is absolute in the sense that it does not depend on anything else for its existence or for a definition of any of its characteristics. On the other hand, just what it is can be defined concretely only through the relationships among the things into which it can be analysed approximately. Each relationship has in it a certain content that is absolute, but this content must, as we have seen, be defined to a closer and closer approximation, with the aid of broader concepts and theories, that take into account more and more of the factors on which this relationship depends. Hence, even though the mode of being of each thing can be defined only relative to other things, we are not led to the point of view of *complete relativity*. For such a point of view implies that there is no objective content to our knowledge at all, either because it is supposed to be defined *entirely* relative to the observer, or to the general point of view and special conditions of each individual, or to special pre-conceptions and modes (or "style") of thinking that may exist in a particular society or in a particular epoch of time.* In our point of view, we admit that all the above things do actually colour and influence our knowledge; but we admit also that nevertheless there still exists an absolute, unique, and objective reality. To know this reality better, and thus to correct and eliminate some of the pre-conceptions and lacunae that are inevitably in our knowledge at any particular time, we must continue our scientific researches, with the objective of finding more and more of the things into which matter in the process of becoming can be analysed approximately, of studying in a better and better approximation the relationships between these things and of discovering in greater and greater detail what are the limitations on the applicability of each specific set of concepts and laws. The essential character of scientific research is, then, that it moves towards the absolute by studying the relative, in its inexhaustible multiplicity and diversity.

* This point of view may perhaps best be characterized by the assumption that "relativity is absolute". In other words, it is stated to be the only absolute truth that there is no absolute content to our knowledge at all.